Pharmacology of Alcohol

Pharmacology of Alcohol

DORA B. GOLDSTEIN, M.D.
Stanford University School of Medicine

New York   Oxford
OXFORD UNIVERSITY PRESS
1983

Library of Congress Cataloging in Publication Data
Goldstein, Dora B.
  Pharmacology of alcohol.
  Bibliography: p.   Includes index.
  1. Alcohol—Physiological effect.   2. Alcohol—
Metabolism.   I. Title.   [DNLM: 1. Alcohol, Ethyl
—Pharmacodynamics.   OV 84 G624p]
QP801.A3G63   1982   615'.7828   82-7939
ISBN 0-19-503111-3   AACR2
ISBN 0-19-503112-1 (pbk.)

Printing (last digit): 9 8 7 6 5 4 3 2 1

Printed in the United States of America

Preface

The drug ethyl alcohol is naturally of interest to pharmacologists because of its pervasive clinical and social presence in our society. It also presents an intellectual challenge in that the simplicity of its chemical structure suggests an unusual mechanism of action. This molecule is unlikely to undergo the specific receptor interactions now known to mediate the effects of many other drugs. Its structural simplicity makes it a feeble agent indeed. That ethanol is a weak drug may surprise those who are concerned about its impact, but a moment's consideration of doses makes it obvious. At nanomolar or micromolar concentrations where other drugs bind strongly to their receptors, ethanol does nothing at all. We take alcohol in tens of grams (or by the ounce) in contrast to other familiar drugs where the usual doses may be a thousand times lower. Its inertness at low concentrations allows its presence in the body to be raised to higher and higher levels, until the simple bulk of it, forced into cell membranes by the pressure of a high aqueous concentration, disrupts the function of cells. Perhaps this is why it has an enormous variety of effects on biochemical systems all over the body. One theory of the action of ethanol ascribes many of its primary effects, especially in the brain, to this disordering of membrane structure. In the liver, the presence of ethanol at high concentrations has other important results. This drug can be metabolized. Indeed, it demands attention from the liver's oxidative systems, competing successfully with endogenous substrates and other sources of energy. The consequences of this are felt throughout the body, as alcohol captures the available oxidizing substances and the redox balance shifts to reduced conditions. These two major mechanisms, the presence of ethanol in membranes and the magnitude of its metabolism in the liver, provide a conceptual framework for a rational approach to the study of alcohol as a pharmacological agent.

This drug interests pharmacologists from many different subdisciplines; one may study it with the techniques of biophysics, biochemistry, physiology, psychology, and pathology, for example. This book grew from a graduate course in which I sought to use ethanol as an illustration of principles and techniques of pharmacology, all applied to a single drug. The book could serve as an introduction to experimental pharmacology, as well as to a clinically important drug. Physicians in clinical training sometimes come back to the pharmacology department requesting more information about alcohol, having now encountered its often dramatic effects in their patients. The medical pharmacology texts may not suffice for their purposes. Multi-volume works such as the series edited by Kissin and Begleiter [1] or the Wallgren and Barry classic [2] are more suitable for specialists or for those seeking more information on specific topics. This brief text may be useful as an introduction.

The disease known as alcoholism encompasses much more than pharmacology, but a knowledge of the drug itself must be the beginning of understanding. It is by no means my intention to cover all medical aspects of alcoholism, much less its many ramifications into the social sciences and the law. We consider here the present state of pharmacological knowledge about interactions of ethanol with membranes, with cells, organs, or sometimes whole animals or people. The book is about experimental pharmacology. Studies that illustrate important principles and useful techniques are described in some detail, because it is by experimentation that the field advances. Naturally I have stressed areas that particularly interest me; the content of the book represents in part a personal choice and in part an attempt to cover the subject matter where credible, well-controlled experiments can be put together into a coherent picture.

Stanford, California   D.B.G.
November 1981

1. Kissin, B. and Begleiter, H., eds. The Biology of Alcoholism. Vol. 1, Biochemistry, 1971. Vol. 2, Physiology and Behavior, 1972. Vol. 3, Clinical Pathology,

1974. Vol. 4, Social Aspects of Alcoholism, 1976. Vol. 5, Treatment and Rehabilitation of the Chronic Alcoholic, 1977. Plenum Press, New York.
2. Wallgren, H. and Barry, H., III. Actions of Alcohol. Vol. 1, Biochemical, Physiological and Psychological Aspects. Vol. 2, Chronic and Clinical Aspects. Elsevier, Amsterdam, 1970.

Contents

Pharmacology of Alcohol

1. Absorption, Distribution, and Elimination of Ethanol

In this first chapter we will consider how the body handles ethanol, beginning with the uncomplicated processes of absorption, distribution, and excretion. Most of the chapter is devoted to ethanol catabolism. Although we are dealing with a very simple chemical reaction indeed, i.e., the oxidation of a two-carbon alcohol to the corresponding aldehyde, there are still many mysteries.

Absorption

Absorption of ethanol from the stomach is simply a matter of diffusion. It is rapid, especially with fairly high alcohol concentrations. Gastric absorption is fastest when strong drinks are taken on an empty stomach. Distilled spirits contain 40–50% ethanol by volume. Dilute beverages, such as beer (3–5% ethanol) or wine (about 12%) are absorbed more slowly and dilution by food also slows absorption. But gastric absorption is not the only process that affects entry of ethanol into the body. The passage of stomach contents into the duodenum is also important because absorption of alcohol from the small intestine is even faster than from the stomach. Alcohol absorption is fastest when the stomach empties quickly. Food taken with the alcohol has a dual effect: it dilutes the ethanol in the gastrointestinal tract and it delays stomach emptying. Distilled beverages, taken on an empty stomach, may cause pylorospasm, so that the alcohol is retained in the stomach and eventually absorbed from there. These are important effects because the rate of

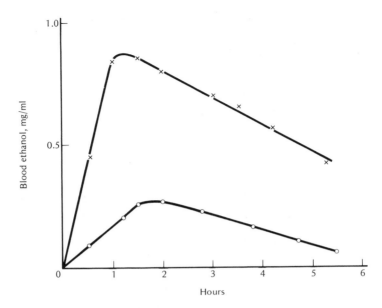

Fig. 1-1. *Effect of food on ethanol absorption.* Blood alcohol in fingertip blood at different times after ingesting alcohol either on an empty stomach (upper curve) or with a meal (lower curve). Data from a single subject. Redrawn from Goldberg [1].

absorption of alcohol determines the levels of alcohol in the blood and the amount that reaches the brain. Figure 1-1 shows blood alcohol curves after a certain dose of ethanol taken orally, either on an empty stomach or with food. In the fasting state, ethanol is absorbed within about an hour and thereafter the curve reflects its elimination. The absorption phase is quite different if food is taken with the ethanol. Not only is the peak later, but it is also much lower. The peak concentration, and thus the maximum degree of intoxication, is much reduced.

Furthermore, taking food with drink sharply reduces the total amount of ethanol that reaches the brain. You will notice that the ethanol is totally eliminated in about the same time in both cases and that the area under these curves, the total exposure of the brain to ethanol under the two conditions, is not the same at all. The explanation is that the ethanol passes through the liver before it reaches the brain or the peripheral blood. During all this time, until the blood alcohol has fallen very low, the liver is working to capacity. The first pass through the liver will dispose of a greater proportion of the dose if the ethanol arrives at a low rather than a high concentration. Consider the extreme case where the same dose of alcohol was taken in tiny sips spread out over a whole day. It is easy to imagine that the liver would handle all of it as fast as it arrived, and none whatsoever would reach the brain. Figure 1-2 shows an experiment where the same amount of ethanol was given to rats in a single dose or in divided doses. Note that all the ethanol was finally eliminated at the same time in both situations; the elimination rate was the same. But the different absorption rates, here achieved by spacing out the doses, produced quite different circulating levels of alcohol.

Distribution

Ethanol rapidly diffuses throughout the aqueous compartments of the body, going wherever water goes, with ease. As must be obvious, it finds the blood-brain barrier no challenge. However, ethanol is not a very lipid-soluble compound. Its partition coefficient between lipid and aqueous phases

Fig. 1-2. *The timing of ethanol intake greatly affects the blood levels.* Ethanol, 4.8 g/kg, was given to rats orally, either in a single dose (upper curve) or as four doses of 1.2 g/kg at 0, 3, 6, and 9 hr (lower curve). From Kalant et al. [2].

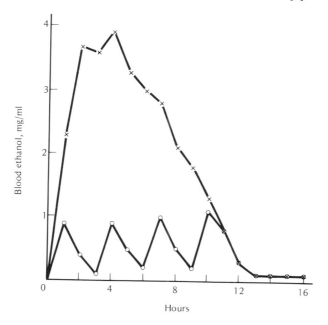

is about 0.1. Concentrations in the body water will always be about ten times higher than in body fats. Unlike some other drugs, alcohol does not accumulate in adipose tissue. It is found in all tissues in a concentration that depends on the water content of the tissue. It can cross the placenta, and damage the fetus.

The volume of distribution is easily calculated from the alcohol elimination curve. The postabsorption rate of elimination is constant with time (see below). Extrapolation back to zero time gives C_0, the concentration that would obtain if the whole dose could be distributed instantaneously. Dividing the dose (g/kg) by C_0 (g/l) gives the volume of distribution as a fraction of the body weight (l/kg). This number, sometimes called Widmark's r, is about 0.8, indicating that ethanol distributes into the total body water. The upper curve of Figure 1-2 extrapolates to a C_0 of about 6 g/l, so the volume of distribution is about 4.8/6, or 80% of body weight.

The rate of distribution of ethanol into different tissues varies according to their blood supply. Brain, which is much more vascular than most other organs, equilibrates rapidly with arterial blood. During absorption, the brain ethanol concentration is much higher than in venous blood, which is still losing ethanol to the tissues that have a smaller blood supply, such as muscle. There will be arteriovenous and brain-venous differences until the whole body has equilibrated. An extreme example of this is shown in Figure 1-3. Here it is obvious that sampling of venous blood will not indicate the degree of intoxication until absorption is complete.

Excretion

Now that we have settled the matter of getting the ethanol into the body, we must consider the more difficult question of how to get it out. A relatively small proportion of the drug, only 5–10%, is excreted unchanged in the expired air and in the urine. This proportion varies with the dose because excretion follows first-order kinetics and metabolism is linear with time. Consequently, at high doses relatively more will be excreted.

Excretion is important for forensic purposes. Samples are often taken for alcohol assay as soon as possible after an accident or crime. If a quantitative answer is needed, blood and breath samples are valid but urine is less reliable. The arterial blood alcohol or, in the postabsorptive phase, the venous blood alcohol obviously represents something close to what the brain is experiencing. Because the lung is an organ well designed to maintain an equilibrium of volatile compounds between blood and air, we can easily

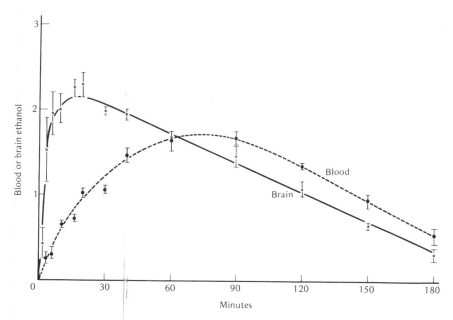

Fig. 1-3. *Tail blood and brain alcohol concentrations in mice.* Mice were given ethanol, 2 g/kg i.p., and five or six mice were used for blood and brain samples at intervals thereafter. Each mouse provided a blood sample at one time point and a brain sample at the next time point. From Goldstein, unpublished.

determine the arterial blood concentration by measuring a sample of alveolar air. Some obvious precautions must be taken, however, such as making sure the mouth is clear of ethanol. The factor 2100 is used to convert breath alcohol to the arterial blood concentration. That is, there is the same amount of ethanol in 1 ml of blood as in 2.1 liters of alveolar air. This high ratio reflects the high water solubility of ethanol.

In contrast to blood and alveolar air, urine gives a crude picture of the level of intoxication. The kidney does indeed excrete the ethanol in a concentration that is the same as the blood (allowing for a slight difference in the water content of the two fluids). But the contents of the bladder represent an integration of the ethanol concentrations that have reached it during whatever time has elapsed since the bladder was last emptied. Only if the recently produced urine is sampled (by emptying the bladder twice) will the urine reflect the current degree of intoxication.

Metabolism

The first step in the metabolism of ethanol is its oxidation to acetaldehyde. This conversion is carried out almost entirely in the liver, although other tissues have some oxidative capacity. The brain's ability to oxidize ethanol is barely detectable. The liver contains three different enzyme systems potentially capable of carrying out this reaction. They are liver alcohol dehydrogenase, catalase, and the microsomal ethanol-oxidizing system. The alcohol dehydrogenase is by far the most important.

Alcohol dehydrogenase. Liver alcohol dehydrogenase is a soluble enzyme that uses nicotinamide adenine dinucleotide (NAD) as coenzyme to convert ethanol to acetaldehyde, as follows:

$$CH_3CH_2OH + NAD^+ \rightleftharpoons CH_3CHO + NADH + H^+$$

This enzyme (which carries the number 1.1.1.1 in the nomenclature of the International Union of Biochemistry) has been extremely well studied by biochemists because it is easily purified, is stable, and shows interesting kinetics. The enzyme contains four atoms of zinc per dimer. The zinc is of biochemical interest but has little practical importance, since no dietary deficiency affects the enzyme, nor are there any useful zinc chelators for use as inhibitors. Theorell and Chance worked out the reaction sequence, wherein the NAD first combines with the enzyme, followed by the ethanol, forming a ternary complex [3]. After the oxidation takes place, the acetaldehyde dissociates. The last step, dissociation of the reduced form of NAD (NADH) from the enzyme protein, is rate-limiting because of the high affinity of NADH for the enzyme active site. It follows that the rate of ethanol oxidation can best be speeded up by addition of substrates that are reduced by liver alcohol dehydrogenase and can oxidize the NADH while it is still attached to the enzyme. A few compounds, including glyceraldehyde, are known to do this [4].

Pyrazole, a simple compound whose structure is shown in Figure 1-4, inhibits alcohol dehydrogenase by combining at the active site after the addition of NAD, forming an inactive ternary complex [5]. The inhibition is competitive with ethanol. Pyrazole is useful for studies of the enzyme action and is sufficiently nontoxic for some *in vivo* use [6]. Derivatives with substitutions in the 4-position, including 4-methylpyrazole, are more potent than the parent compound. Pyrazole is the basis for an affinity chromatography technique for purification of alcohol dehydrogenases [7]. A Se-

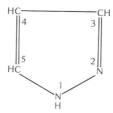

Fig. 1-4. *Pyrazole.*

pharose column, to which a pyrazole derivative is attached, retains the pyrazole-sensitive alcohol dehydrogenase, applied as a relatively crude preparation from liver, in the presence of NAD. The enzyme can be eluted with ethanol.

In vivo, ethanol elimination is generally observed to proceed at a constant rate, once the absorption phase is over. Zero-order conditions may be interpreted as Michaelis-Menten kinetics with the enzyme saturated. Indeed, it is saturated most of the time, during alcohol elimination. The human liver alcohol dehydrogenase has a K_m for ethanol of about 2 mM, which is below intoxicating levels. Therefore, it does seem reasonable that the amount of enzyme determines the linear rate of alcohol elimination at all but the lowest ethanol concentrations. But if this were so, alcohol elimination rates should agree with liver alcohol dehydrogenase activity among individuals of different ages, sex, drug histories, or species. There is, in fact, no such correlation between enzyme activity, measured *in vitro* under optimal conditions, and rates of ethanol elimination *in vivo*. From this we can conclude that even though the enzyme is saturated with substrate, something else is limiting the rate of the reaction. It is the other reactant, NAD, which must be continuously regenerated from NADH as the latter is formed during ethanol oxidation. This brings us to a most important point about ethanol metabolism, its magnitude. Ethanol doses are far in excess of the doses of other drugs. They divert the liver function so that many normal processes come to a halt while the liver deals with the ethanol. For example, a fairly heavy evening's drinking could amount to about 100 g of ethanol (about six strong drinks). This is just over 2 moles. It will take 2 moles of NAD to oxidize it by means of the liver alcohol dehydrogenase pathway. Two moles of NAD weigh a kilogram and a half! Clearly the liver must continuously regenerate its much smaller supply of NAD. This occurs by means of a variety of NAD-dependent dehydrogenases, and the overall rate of this process must be determined by the activity of these enzymes and the availability of the oxidized forms of their substrates. It is not surprising that these factors would vary widely among individuals or in different nutritional states. Fasting reduces the rate of elimination of ethanol, presumably by reducing the supply of substrates for dehydrogenases [8]. Alcoholics know of this effect and may reduce their food intake to obtain the maximum effect from the available liquor.

The wholesale conversion of NAD to NADH during ethanol oxidation overwhelms the liver. The regeneration of NAD cannot quite keep up, and there is an increase in the ratio of NADH to NAD. In equilibrium with this pair of coenzymes are the substrate pairs for other dehydrogenases, and their ratios too will be found to shift in the reduced direction. The ratio of lactate to pyruvate rises, and this provides a convenient estimate of the redox balance [9]. Figure 1-5 illustrates this ratio in an *in vitro* experiment.

A similar shift is found inside the mitochondria, where the reoxidation of the NADH takes place. Since NADH itself does not cross the mitochondrial membrane, the process is accomplished by substrates that do. For example, the excess NADH in the cytosol reduces some oxalacetate to malate. The latter can cross into the mitochondrion where it will be reoxidized by the respiratory chain and emerge again into the cytoplasm in the form of oxalacetate to take up another molecule of NADH. This process affects the redox balance inside the mitochondrion. The Krebs cycle is slowed or actually stopped, partly by excess of NADH. It has been shown that most of the oxygen taken up by the liver during ethanol metabolism can be accounted for by oxidation of the ethanol itself [9]. Thus it is easy to see why metabolism of other compounds, such as fatty acids, is severely disrupted during this time.

Most efforts to find a sobering-up agent are aimed at the dehydrogenase action. Compounds that can be reduced by NADH should be helpful, particularly if they are substrates for the alcohol dehydrogenase itself and can react with the NADH while it is still in combination with the enzyme. The only agent that has been at all successful is fructose, which may act in two ways. It can be directly reduced to sorbitol with an NADH-requiring enzyme. More importantly, it can be metabolized to glyceraldehyde, which is a substrate for alcohol dehydrogenase and can regenerate NAD on the enzyme molecule.

Fructose, however, has not proven to be very useful in a practical sense. In some studies, no effect could be found *in vivo*, and even where there

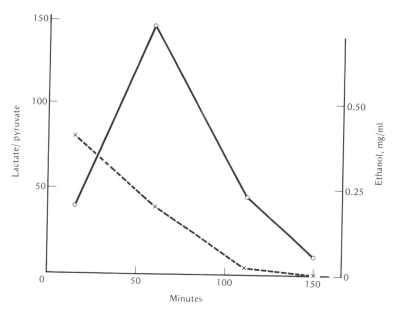

Fig. 1-5. *Liver redox balance.* The lactate/pyruvate ratio (O—O) increased in rat liver slices during oxidation of ethanol. Alcohol was added at 15 minutes; its concentration is shown by the broken line. From Forsander [9].

was an increase in ethanol elimination, it was of small magnitude—not enough to be clinically helpful. An increase of 20–30% in elimination rate will not appreciably improve the condition of the dangerously intoxicated patient in the emergency room. Several hours would still elapse before the blood alcohol returned to harmless levels. In any case, if one did succeed in greatly increasing the rate of ethanol oxidation, it would then be necessary to deal with a flood of acetaldehyde, the product of the reaction. Furthermore, because this is a stoichiometric reaction between the added substrate and NAD, the doses of the sobering-up agent must be of about the same magnitude as the ethanol doses. A teaspoon of honey will not help at all.

The *initial* rate of ethanol elimination may not be limited by availability of NAD. This rate can only be measured by estimating the loss of ethanol from the whole body (in mice, for example [10]), since the ethanol levels in the blood at early times reflect simultaneous absorption and elimination. The initial rapid rate probably represents the action of liver alcohol dehy-

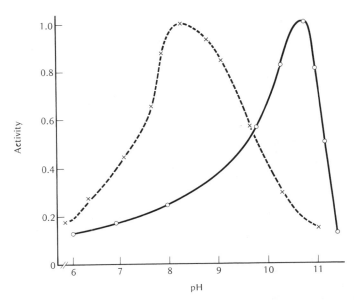

Fig. 1-6. *pH activity curves of alcohol dehydrogenases purified from two human livers.* The normal enzyme (O—O) peaks at a much higher pH than the atypical enzyme (×—×). Redrawn from Von Wartburg et al. [11].

drogenase unimpeded by a shortage of NAD, before the redox balance has been upset. This implies that there will be a higher acetaldehyde level during the early stages than later, and that different activity of alcohol dehydrogenase in different individuals may be expressed temporarily before NAD regeneration becomes rate-limiting.

Liver alcohol dehydrogenase exists in several genetically determined variants. Some individuals have an "atypical" form of the enzyme [11], with unusually high specific activity at physiological pH. While the "usual" enzyme has a pH optimum around 11, the newly discovered atypical form works best at pH 8.5. Figure 1-6 shows this comparison. The variant form also reacts differently with inhibitors; it is inhibited by thiourea rather than activated as is the normal enzyme. It can thus be easily identified in biopsy or autopsy material. At first it was surprising to find that people who carried the superactive enzyme did not metabolize ethanol faster than others. This can now be explained by the fact that the enzyme activity does not control the elimination rate. Genetic polymorphism with respect to liver alcohol dehydrogenase and its possible relation to dysphoric reactions to ethanol among Oriental people will be discussed in Chapter 10.

Recently, yet another variant of liver alcohol dehydrogenase has been described [12]. This isozyme, π-alcohol dehydrogenase, is unique in its high K_m for ethanol (20 mM) and its insensitivity to inhibition by pyrazole. It is readily separated from pyrazole-sensitive alcohol dehydrogenases by the affinity chromatography method mentioned above, since it is not retained by the column. This isozyme can contribute substantially to ethanol oxidation at high concentrations *in vitro*, but has not yet been shown to affect elimination rates *in vivo*, or to have any special relation to alcoholism.

Catalase. A second enzyme that can oxidize ethanol is catalase, which abounds in liver, apparently for protection against hydrogen peroxide formed by various oxidases. Catalase can act as a peroxidase, using hydrogen peroxide to oxidize other substances. Ethanol is a good substrate for this reaction *in vitro* [13]. However, *in vivo* the reaction seems to be of little quantitative importance. One cannot affect the rate of alcohol elimination by inhibiting catalase, even when the enzyme activity is reduced to 10% of normal. However, this is an inconclusive experiment because the liver has a huge excess of catalase. Again, as with the dehydrogenase, it is likely that the coreactant is rate-limiting. The supply of hydrogen peroxide is probably not sufficient for oxidation of large amounts of ethanol. Hydrogen peroxide is formed by several oxidases in the liver, but combinations of enzymes and substrates are probably not present in sufficient amounts to drive the ethanol oxidation with peroxide.

The best evidence against a significant role of catalase is that ethanol oxidation in rats was shown to be unaffected by catalase inhibition under conditions where methanol oxidation was inhibited [14]. As shown in Figure 1-7, addition of a peroxide-generating system stimulated methanol but not ethanol oxidation in rat liver slices. These experiments, the only convincing ones on the catalase question, show that catalase plays no significant role in ethanol oxidation, at least in the rat.

In humans, methanol is apparently oxidized both by catalase and by liver alcohol dehydrogenase. Ethanol competes with methanol for sites on either enzyme. This is the basis for treatment of methanol poisoning by administration of ethanol. Methanol is not a very toxic drug, less so than ethanol, but is oxidized to formaldehyde, which is very reactive and harmful. The presence of alcohol dehydrogenase in the retina (where it participates in

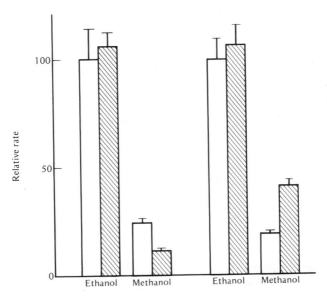

Fig. 1-7. *Role of catalase in oxidation of methanol and ethanol by rat liver slices.*
The rate was measured as conversion of ^{14}C-alcohols to aldehydes and trapping with
semicarbazide. Left, treatment of the rats with aminotriazole, a catalase inhibitor
(hatched bars) partially blocked methanol but not ethanol oxidation. Right, addition
of xanthine (hatched bars) to the incubation medium stimulated methanol but not
ethanol oxidation. Xanthine, along with endogenous xanthine oxidase, provided a
hydrogen peroxide generating system. Data from Smith [14].

metabolism of visual pigments) may account for the specific retinal toxicity
of methanol. Administration of ethanol retards the metabolism of methanol
so that most of the dose can be excreted harmlessly in the breath and
urine. An analogous situation occurs with ethylene glycol; its oxidation
product, oxalic acid, is toxic to the kidney and its metabolism by alcohol
dehydrogenase can be blocked by ethanol.

Microsomal ethanol-oxidizing system (MEOS). Finally, a third mechanism
is potentially at work. Microsomal fractions of liver can oxidize ethanol in
the presence of molecular oxygen and NADPH (the reduced form of nicotinamide adenine dinucleotide phosphate). This resembles the general
mixed-function oxygenase system that handles a variety of exogenous compounds. Whether the microsomal ethanol-oxidizing system (MEOS) is
identical with the mixed-function oxidase is still controversial. Ethanol binds
to cytochrome P-450, as do other substrates for liver microsomes; however,

ligands are not necessarily substrates. Ethanol oxidation is inhibited by one well-known microsomal inhibitor, carbon monoxide, but not by another, SKF 525A. Ethanol is not unique among substrates for the microsomal drug-oxidizing system in failing to respond to this agent. Lieber and colleagues, who have characterized this system, believe it is responsible for a substantial part of the ethanol elimination in the liver, especially at high blood alcohol levels. The pH optimum of MEOS is in the physiological range and its K_m for ethanol is high—about 8 mM. These investigators also credit MEOS with responsibility for increased rates of ethanol elimination after chronic administration. In rats, chronic feeding of ethanol increases MEOS activity, hepatic microsomal protein, smooth endoplasmic reticulum, and cytochrome P-450. This clearly resembles the induction of a mixed-function oxidase system. The existence of this system may explain the inhibition by ethanol of metabolism of other drugs handled by microsomal enzymes, as well as metabolic cross-tolerance with some such drugs.

Early objections to these claims were based on possible contamination of the system with catalase. The liver catalase exists in small particles called peroxisomes that are intermediate in size between mitochondria and microsomes, and thus catalase commonly contaminates microsomal preparations. An NADPH oxidase in these fractions is a potential source of hydrogen peroxide. By using inhibitors of catalase and of alcohol dehydrogenase (not a likely contaminant), Lieber and DeCarli [15] were able to show that the microsomes contained ethanol-oxidizing activity that could not be accounted for by the other enzymes. It remains to be shown how much this route of oxidation contributes to ethanol elimination. This is extremely difficult, because one cannot easily compare *in vivo* and *in vitro* conditions. Ethanol elimination mediated by alcohol dehydrogenase, as pointed out above, seems to depend on the activity of a number of other dehydrogenases that regenerate NAD, and on the concentrations of their substrates. Thus the activity measured *in vitro* under optimal conditions has little to do with *in vivo* rates. The controversy continues, and the difficulty of accounting for *in vivo* elimination rates by particular enzymes is compounded by the recent discovery of π-alcohol dehydrogenase (described above) which, like MEOS, has a high K_m and is insensitive to pyrazole.

Overall ethanol elimination rate

The rate at which ethanol disappears from the blood varies widely among individuals. A general rule is useful, even if crude. For this purpose, one can use the figure of 10 ml of absolute alcohol per hour for an average-size man. Small people, women, and teenagers may be expected to eliminate ethanol more slowly, in general. A "drink" usually contains 15–20 ml of alcohol. There is approximately the same amount of ethanol in a 12-oz. can of beer (4.5% alcohol by volume), a 4-oz. glass of wine (12%), and a scant jigger (1.2 oz.) of whiskey (45%).

Summary

Gastric absorption of alcohol is slowed by the presence of food in the stomach, which substantially affects the amount of ethanol reaching the brain. Some ethanol is excreted unchanged in the breath and the urine, but most is eliminated by oxidation in the liver. Alcohol dehydrogenase is the main enzymatic pathway, and the rate-limiting process is probably the regeneration of NAD for the dehydrogenase. The great need for NAD to handle a large load of ethanol may force the liver to shortchange other NAD-requiring reactions, resulting in a significant shift in the redox balance with an increase in reduced intermediates, such as lactate. The metabolism of carbohydrates, fats, and hormones is affected. Several isozymes of alcohol dehydrogenase are known and their relevance to alcoholism is being investigated. The liver microsomal ethanol-metabolizing system (MEOS) may also participate in elimination of ethanol. It may play a role in metabolic drug interactions, but we do not yet know what proportion of ethanol elimination can be ascribed to it.

References

1. Goldberg, L. Quantitative studies on alcohol tolerance in man. Acta Physiol. Scand. 5: suppl. 16, 1943.
2. Kalant, H., Khanna, J.M., Seymour, F. and Loth, J. Acute alcoholic fatty liver—metabolism or stress. Biochem. Pharmacol. 24: 431–434, 1975.
3. Theorell, H. and McKinley-McGee, J.S. Liver alcohol dehydrogenase. I. Kinetics and equilibria without inhibitors. Acta Chem. Scand. 15: 1797–1810, 1961.
4. Holzer, H. and Schneider, S. Zum Mechanismus der Beeinflussung der Alkoholoxydation in der Leber durch Fructose. Klin. Wochenschr. 33: 1006–1009, 1935.
5. Theorell, H., Yonetani, T. and Sjöberg, B. On the effects of some heterocyclic compounds on the enzymic activity of liver alcohol dehydrogenase. Acta Chem. Scand. 23: 255–260, 1969.
6. Goldberg, L. and Rydberg, U. Inhibition of ethanol metabolism in vivo by administration of pyrazole. Biochem. Pharmacol. 18: 1749–1762, 1969.

7. Lange, L.G. and Vallee, B.L. Double-ternary complex affinity chromatography: preparation of alcohol dehydrogenases. Biochemistry 15: 4681–4686, 1976.
8. Smith, M.E. and Newman, H.W. The rate of ethanol metabolism in fed and fasting animals. J. Biol. Chem. 234: 1544–1549, 1959.
9. Forsander, O.A. Influence of the metabolism of ethanol on the lactate/pyruvate ratio of rat-liver slices. Biochem. J. 98: 244–247, 1966.
10. Forney, R.B., Hughes, F.W., Hulpieu, H.R. and Clark, W.C. Rapid early metabolism of ethanol in the mouse. Toxicol. Appl. Pharmacol. 4: 253–256, 1962.
11. Von Wartburg, J.-P. Papenberg, J. and Aebi, H. An atypical human alcohol dehydrogenase. Can. J. Biochem. 43: 889–898, 1965.
12. Li, T.-K., Bosron, W.F., Dafeldecker, W.P., Lange, L.G. and Vallee, B.L. Isolation of π-alcohol dehydrogenase of human liver; Is it a determinant of alcoholism? Proc. Nat. Acad. Sci. 74: 4378–4381, 1977.
13. Keilin, D. and Hartree, E.F. Properties of catalase. Catalysis of coupled oxidation of alcohols. Biochem. J. 39: 293–301, 1945.
14. Smith, M.E. Interrelations in ethanol and methanol metabolism. J. Pharmacol. Exp. Ther. 134: 233–237, 1961.
15. Lieber, C.S. and DeCarli, L.M. Hepatic microsomal ethanol-oxidizing system. In vitro characteristics and adaptive properties in vivo. J. Biol. Chem. 245: 2505–2512, 1970.

Review

Khanna, J.M. and Israel, Y. Ethanol metabolism. Internat. Rev. Physiol. 21: 275–315, 1980.

2. Acetaldehyde

Acetaldehyde has several interesting pharmacological effects of its own and a possible, much debated, role in the actions of ethanol. Since acetaldehyde is a highly reactive compound and a much more potent drug than ethanol, one naturally wonders whether the effects of ethanol are mediated by acetaldehyde. Evidence presented in this chapter indicates that acetaldehyde is not the cause of acute intoxication, but is responsible for the disulfiram reaction. We will also consider how acetaldehyde can participate in some interesting condensation reactions with catecholamines, and what their importance might be.

Acetaldehyde in blood and tissues

All the reactions discussed in the preceding chapter have accomplished only the first step of ethanol oxidation, its conversion to acetaldehyde. This product is further oxidized by several hepatic enzymes, aldehyde oxidases and dehydrogenases; the most important of these is a low-K_m mitochondrial aldehyde dehydrogenase [1]. Most of the acetaldehyde that is formed by ethanol oxidation is swept up immediately by these enzymes and converted to acetate, a relatively harmless compound. The acetate cannot be further metabolized by the liver if ethanol is being metabolized at the same time because the Krebs cycle is stalled due to the abnormal hepatic redox balance. Acetate therefore spills out into the blood where it is easily taken care of by other tissues, particularly skeletal muscle. Thus the ethanol is finally converted to carbon dioxide.

Since the aldehyde dehydrogenases in the liver cannot quite keep up with acetaldehyde production, small amounts of acetaldehyde escape to circulate in the blood stream during oxidation of ethanol. Gas chromatography is the standard technique for determining acetaldehyde in biological materials. Usually the head space over an aqueous solution is sampled; this contains only the volatile components of the blood or tissue. Ethanol can be measured simultaneously. Unfortunately, assays of acetaldehyde in biological samples are subject to artifacts in both directions; acetaldehyde may be lost or it may be formed nonenzymatically from ethanol. There are species and tissue differences in the extent of spurious formation or loss of acetaldehyde. Reports of high blood acetaldehyde concentrations (e.g., 10 μg/ml) after ingestion of ethanol are suspect, particularly if the acetaldehyde concentration varies with that of ethanol.

The concentration of acetaldehyde in blood should remain stable during metabolism of ethanol, because the rate of ethanol oxidation (and thus of acetaldehyde formation) is constant. Steady state conditions should prevail and there should be a constant level of acetaldehyde in the blood until blood alcohol levels have fallen to the low range where first-order elimination takes place. Figure 2-1 shows experimental confirmation of this expectation.

The circulating acetaldehyde does not necessarily enter the brain. Little or no acetaldehyde was found in rat brain after ethanol administration, when appropriate corrections were made for the acetaldehyde content of the cerebral blood [3]. When acetaldehyde itself was injected, some appeared in brain, but only when the blood acetaldehyde levels were very high, above 10 μg/ml (see Fig. 2-2). The barrier to acetaldehyde probably consists of aldehyde-metabolizing enzymes in the cerebral endothelium.

Role in acute intoxication

The evidence that acetaldehyde does not directly mediate ethanol intoxication is strong, although it is mostly circumstantial. For one thing, the behavioral signs of intoxication increase and later recede as the blood alcohol levels rise and fall (Fig. 2-3) [4]. During this time the blood acetaldehyde levels are quite stable (Fig. 2-1), so that the behavior correlates with concentrations of ethanol, not with those of acetaldehyde.

A stronger piece of evidence is that pyrazole does not prevent intoxication. Indeed, at a given blood level of ethanol, pyrazole-treated animals are

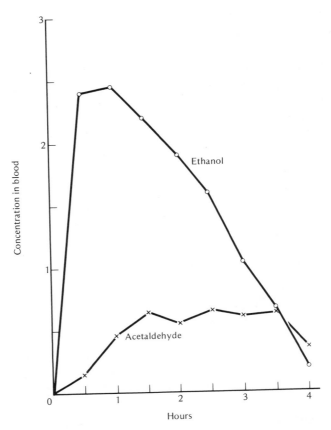

Fig. 2-1. *Stability of blood acetaldehyde levels during oxidation of ethanol.* Rats were given ethanol, 2 g/kg, by gavage. Blood levels of ethanol (O—O) are in mg/ml and those of acetaldehyde (×—×) in µg/ml. From Kalant et al. [2].

more impaired than controls. If acetaldehyde were causing the intoxication, the pyrazole-treated rats should be sober.

In vitro, aldehydes can block stimulated respiration of brain slices [5]. Alcohols also have this effect, but it is not mediated by their conversion to aldehydes. When rat or guinea pig brain slices are stimulated, either electrically or by adding high concentrations of potassium, their oxygen uptake is increased. The extra respiration is blocked by ethanol and by other CNS depressant drugs, such as barbiturates and opiates. In this reaction, as in many other actions of alcohols to be discussed below, the potency of aliphatic alcohols increases sharply with their chain length. Aldehydes also inhibit the stimulated respiration of brain slices, and acetaldehyde is about 200 times more potent than ethanol in this regard. However, the longer chain aldehydes are no more potent than the short-chain aldehydes, quite unlike the alcohols. The longer-chain alcohols could not be metabolized rapidly enough to produce effective concentrations of aldehydes, so it is unlikely that the alcohols work in this biochemical system by being converted to aldehydes.

Sympathomimetic effects

Effects of aliphatic aldehydes on the sympathetic nervous system were described by Eade in 1959 [6]. Injection of acetaldehyde caused contraction

Fig. 2-2. *Acetaldehyde appears in brain only at extremely high blood acetaldehyde levels.* The points represent acetaldehyde concentrations in cerebral blood and in brain of rats given ethanol (2 g/kg, i.p.) and different doses of acetaldehyde (50–150 mg/kg i.p.). The points at the lower left, under 0.5 μg/g brain, were at the limit of sensitivity of the assay method. From Sippel and Eriksson [3].

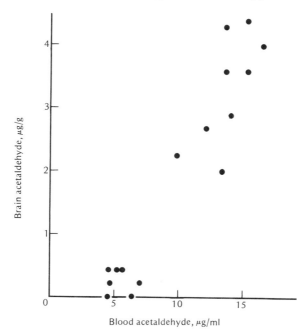

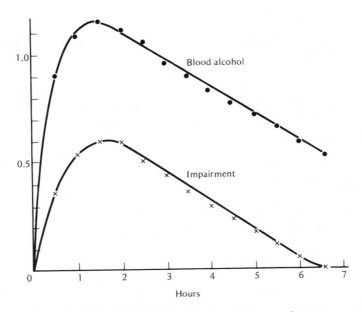

Fig. 2-3. *Temporal correlation of blood alcohol and behavioral impairment in man.*
The points are means for 17 experiments where the impairment of motor coordination was measured repeatedly as the blood alcohol rose and fell. The blood alcohol units are mg/ml; motor incoordination in the finger-finger test (described in Chapter 6) is shown in arbitrary units. From Goldberg [4].

of the nictitating membrane and a brief rise in blood pressure in spinal cats. Either epinephrine or norepinephrine could cause the same response, and so could propionaldehyde and butyraldehyde. The effect was not prevented by administration of a ganglionic blocking agent, hexamethonium. Cocaine, which magnifies norepinephrine's actions by preventing its recapture by the cell that released it, potentiated the effects of aldehydes and catecholamines on blood pressure and on the nictitating membrane. Acetaldehyde had no effect in reserpinized animals, whose norepinephrine stores were depleted. Acetaldehyde clearly works, at least in part, by release of norepinephrine from nerve terminals. Several other studies have confirmed the pressor action of acetaldehyde.

James and Bear [7] demonstrated positive chronotropic and inotropic effects of acetaldehyde on the heart. Using anesthetized dogs, they infused high concentrations of acetaldehyde into the sinus node artery, a branch of the right coronary. The effect was blocked by propranolol and by prior treatment with reserpine, indicating that it was mediated by norepinephrine release at β-adrenergic receptor sites. Related compounds such as ethanol, methanol, and formaldehyde had no effect except at very high concentrations, and then they had negative inotropic effects.

Acetaldehyde is a component of cigarette smoke and was studied in that connection by Egle [8], who administered it to rats by inhalation for brief periods. Again there was a pressor effect and a slight increase in heart rate, but the author concluded that there would be no such action at concentrations found in cigarette smoke.

Disulfiram

The discovery that disulfiram (tetraethylthiuram disulfide, Antabuse) causes adverse reactions after ethanol drinking was serendipitous. In 1948, Hald and Jacobsen were testing the drug for its antihelminthic actions. Taking some themselves, as old-fashioned pharmacologists, they observed that they felt quite sick after drinking. This experience eventually led to the use of disulfiram as a deterrent to drinking, under the trade name Antabuse (Fig. 2-4).

Disulfiram can be taken daily with no discernible effects, as long as one does not drink. However, alcohol ingestion produces an alarming reaction [9]. There is a sensation of heat in the face, readily visible as an intense flushing of the face and neck, sometimes spreading to the chest and arms. At the same time, palpitations are felt and the heart rate increases, at first with no change in blood pressure. There is some dyspnea and perhaps hyperventilation, and nausea and vomiting may occur. After a while, the flush disappears and at that point there may be a sharp fall in blood pressure. Not surprisingly, an intense feeling of uneasiness is also reported.

The syndrome is ascribed to acetaldehyde. Using rabbits and human subjects, the original investigators showed that blood acetaldehyde levels were much higher during ethanol metabolism if disulfiram had been ad-

Fig. 2-4. *Disulfiram.*

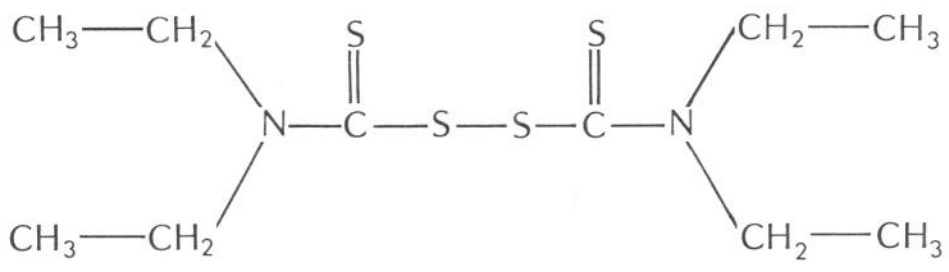

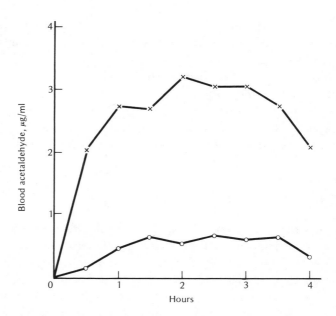

Fig. 2-5. *Disulfiram raises acetaldehyde levels after alcohol administration.* Rats were given ethanol, 2 g/kg, by gavage. One group of animals (upper curve) had been treated with disulfiram; those represented by the lower curve had not. The blood ethanol levels were not affected by disulfiram. From Kalant et al. [2].

ministered the day before. Figure 2-5 shows this in a recent experiment with rats.

Furthermore, when acetaldehyde itself was administered to humans, it caused a facial flush, hyperventilation, and tachycardia [10]. Disulfiram raises blood acetaldehyde levels by inhibiting several different enzymes that oxidize acetaldehyde. The low-K_m aldehyde dehydrogenase of liver mitochondria is inhibited by disulfiram, as is aldehyde oxidase and even xanthine oxidase, which can also oxidize acetaldehyde. Disulfiram has active metabolites, including diethyldithiocarbamate.

Incidentally, the Danish disulfiram experiments showed that no appreciable acetaldehyde was circulating when disulfiram was administered alone [11]. Thus, there is no endogenous formation of acetaldehyde in the absence of ethanol metabolism.

There have been objections to the acetaldehyde hypothesis of disulfiram action. Animals do not always show the signs seen in human subjects, and the severity of alcohol-disulfiram reactions in humans does not always correlate well with the levels of acetaldehyde in the blood. The main difficulty is in explaining the hypotension of the disulfiram-alcohol reaction, which can be severe and is opposite to the cardiovascular effects of acetaldehyde itself. Most of the discrepancy is probably due to the fact that it is difficult to maintain blood concentrations of acetaldehyde by administering acetaldehyde itself, because it is so rapidly metabolized. In the experimental situation, there may not be time for secondary effects, such as the delayed hypotension.

It is now known that disulfiram has another effect that was not suspected at the time of the Jabobsen-Hald experiments. It inhibits dopamine β-oxidase, the enzyme that converts dopamine to norepinephrine. Disulfiram chelates copper, which is a necessary component of the enzyme. The effect on norepinephrine synthesis is sometimes used in experimental studies of catecholamine metabolism in animals or *in vitro*, but is apparently not important in the clinical use of disulfiram.

Disulfiram is used in long-term treatment of alcoholics. A daily dose produces no known effects, but alcohol consumption is followed by the alarming reaction described above. Rarely, this can be severe—a few deaths due to cardiovascular collapse have even been reported. Like many other treatments for alcoholism, this one works in some patients and not in others, according to motivation. Failure is often due to the patients' staying sober for a few months and then stopping the disulfiram medication in the belief that they have recovered from their alcohol problem. Drinking is likely to resume thereafter. Attempts have been made to develop ways of giving disulfiram in long-lasting form. The implanted dose, which is supposed to last for six months, is often only 1 g—equivalent to a few days' oral doses. Furthermore, the extremely low solubility of the drug prevents its absorption from the intramuscular site. Most of it may still be in place after a few months, during which time plasma levels of the drug are below the effective range. A recent double-blind study [12] with twenty alcoholic subjects showed that there was no response to ethanol challenge five days after implantation and no measurable disulfiram in blood at that time. Nevertheless, during the next eight months, seven of the ten disulfiram-implanted subjects and none of the ten placebo controls had a disulfiram-ethanol reaction after drinking. The reactions were generally mild but prolonged. They did not necessarily occur when drinking began, but were preceded by binges of several days' duration. The authors suggest that ethanol may facilitate the absorption of disulfiram from the implantation site, which would explain some puzzles. The disulfiram-treated subjects in this study were

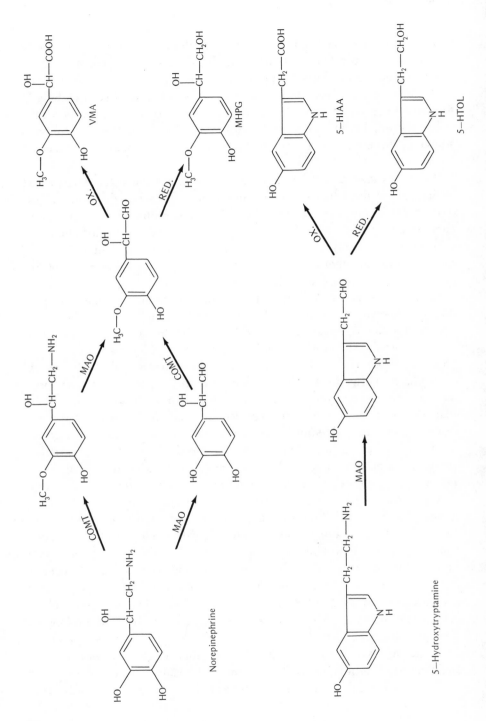

abstinent significantly more than the controls during a two-year follow-up period.

Condensation products, tetrahydroisoquinolines (TIQs)

Acetaldehyde reacts readily with many other compounds. In doing so, it might form pharmacologically active substances, the most interesting of which are the condensation products of acetaldehyde with certain of the neurotransmitters. This topic provides one of the liveliest controversies in alcohol research today. The original observation was that urinary concentrations of the major metabolites of norepinephrine and serotonin were decreased after ethanol administration. At first it was thought that the alcohol had slowed down the turnover of the neurotransmitters, but soon it was discovered that the amines were represented in the urine by their reduced metabolites, rather than the usual acids. These amines are metabolized by O-methyltransferases and monoamine oxidase as shown in Figure 2-6. The aldehyde that is the immediate product of monoamine oxidase action can be further metabolized in either of two ways. Peripherally, the norepinephrine-derived aldehyde is usually oxidized to vanillylmandelic acid (3-methoxy-4-hydroxymandelic acid, VMA) and the serotonin aldehyde is converted to 5-hydroxyindoleacetic acid. But the aldehydes can also be reduced, and in fact, this is the major pathway of norepinephrine metabolism in the brain. During alcohol metabolism, the redox balance of the whole body is shifted toward reducing conditions, as described in Chapter 1. More NADH is available to favor the reduction and less NAD for oxidation. Further, acetaldehyde inhibits the oxidation of the aldehydes formed by monoamine oxidase. Reduced products predominate; i.e., norepinephrine is reduced to 3-methoxy-4-hydroxyphenylglycol and serotonin to 5-hydroxytryptophol. These compounds were found not to have pharmacological effects at any reasonable concentrations, but this was the beginning of a long story.

Fig. 2-6. *Pathways of norepinephrine and serotonin metabolism.* Norepinephrine is catabolized by catechol-0-methyltransferase (COMT) and monoamine oxidase (MAO), acting sequentially in either order. The aldehyde thus formed may either be oxidized to vanillylmandelic acid (VMA) or reduced to 3-methoxy-4-hydroxy-phenylglycol (MHPG). The pathways for serotonin (5-hydroxytryptamine) are analogous except that no methyltransferase is involved. The products are 5-hydroxyindoleacetic acid (5-HIAA) or 5-hydroxytryptophol (5-HTOL). During the metabolism of ethanol the reductive pathways are favored.

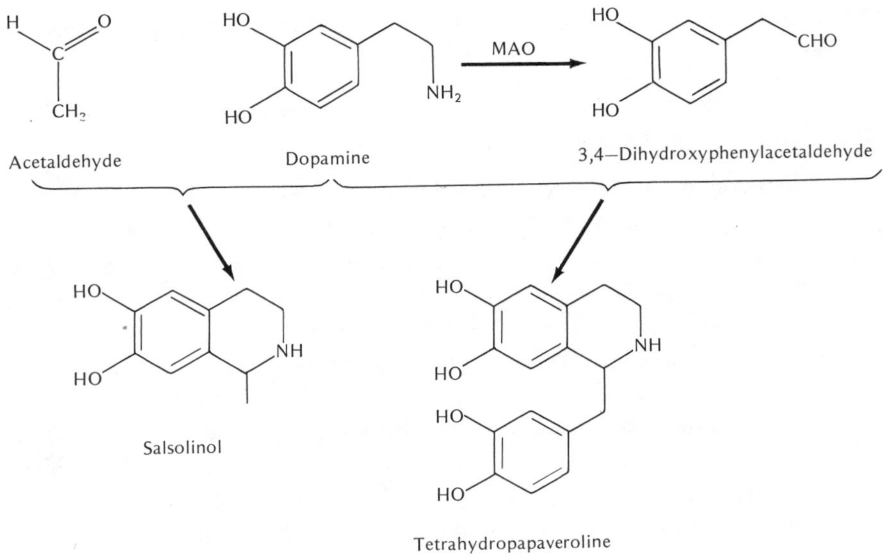

Fig. 2-7. *Formation of two tetrahydroisoquinolines, salsolinol and THP, from do-pamine.*

When the analogous products of dopamine metabolism were sought, the trail took a new turn. Davis and Walsh [13] incubated brain stem homo-genates *in vitro* with labeled dopamine and found a substantial amount of tetrahydropapaveroline (THP), a condensation product of dopamine with its own aldehyde, formed by the action of monoamine oxidase (MAO). Fig-ure 2-7 shows this reaction. (This condensation had been known previ-ously, but had attracted little attention.) Addition of ethanol or acetalde-hyde to the *in vitro* system increased the amount of THP, presumably because further metabolism of the dopamine aldehyde had been blocked.

THP is of particular interest because it occurs in the opium poppy as a precursor of morphine. This led Davis and Walsh to suggest a common basis for opiate and alcohol addiction, linked by THP after its presumed conversion to an opiate in the brain. Other scientists objected to this incau-tious speculation, noting that the clinical syndromes of alcoholism and mor-phine addiction are quite distinct. Furthermore, it was shown experimen-tally that alcohol-dependent mice did not react to the opiate antagonist naloxone the way morphine-dependent mice do [14]. Dependence on ethanol is not the same phenomenon as dependence on morphine.

Other condensation products also have been studied. Organic chemists use the Pictet-Spengler reaction to condense phenylethylamines with aldehydes at high temperature and low pH. This reaction proceeds under physiological conditions, nonenzymatically, if the phenylethylamine is hydroxylated. Since catecholamines are hydroxylated, they will react. (A similar reaction is the basis for the well-known fluorescence histochemistry of the biogenic amines. A condensation occurs in tissue slices exposed to formaldehyde vapor, and a further oxidative step forms the fluorescent compound *in situ*.) Thus, acetaldehyde can condense nonenzymatically with catecholamines to form TIQs and with indoleamines to form β-carbolines. A Pictet-Spengler reaction of dopamine and acetaldehyde to form a compound known as salsolinol is shown in Figure 2-7. Cohen and co-workers [15] have gathered evidence that TIQs might act as false transmitters in adrenergic systems. The compounds can be taken up into catecholamine storage granules and released by the same stimulus that would release catecholamine. Some typical adrenergic effects were seen, e.g., mydriasis and exophthalmia after uptake into the nerve terminals of the rat eye and subsequent stimulation. The response is thus ascribed to release and pharmacological action of the TIQ. These compounds can also inhibit dopamine-sensitive adenylate cyclase in rat striatum.

This evidence is relevant to ethanol's actions only if such compounds exist in brain *in vivo* after ethanol administration. Collins and Bigdeli [16] found salsolinol, the dopamine-acetaldehyde condensation product (Fig. 2-7), in brains of rats after treatment with ethanol and pyrogallol, an inhibitor of catechol-O-methyltransferase. This inhibition should prevent metabolism of TIQs by O-methylation. They found about 17 ng of salsolinol per gram of brain, which is about 1% of the dopamine content of whole brain. When the rats were further treated with pargyline, the amounts of apparent salsolinol were as high as 120 ng/g. This effect was ascribable to inhibition of acetaldehyde metabolism by pargyline, a previously unknown effect of the drug. Turner et al. [17] have also found a condensation product in the brain of rats. They administered L-DOPA along with an inhibitor of peripheral DOPA decarboxylase (to increase the availability of DOPA in the brain) in the drinking water for eight days. They identified small amounts of THP in brain, and slightly more was seen if ethanol was administered in addition to the L-DOPA and decarboxylase inhibitor.

These alkaloids have been found in brain only when catecholamine metabolism has been interfered with. No one has found them after administration of ethanol alone. Even an attempt to develop the compounds by chronic administration of ethanol failed to reveal them. O'Niell and Rah-wan [18] looked for salsolinol in brains of rats after several days' adminis-tration of ethanol by inhalation. They found none, although their assay would have picked up as little as 8 ng/g. The rats were shown to have had effec-tive amounts of ethanol, since their blood ethanol levels went as high as 5 mg/ml and they showed a clear ethanol withdrawal reaction.

Recently, a TIQ metabolite, O-methylsalsolinol, has been found in brains of mice after chronic administration of ethanol [19]. This compound may have a longer biological lifetime than the parent TIQ; its pharmacological activity is unknown.

There was little or no TIQ in the brain and it was known that no phar-macological effects could be seen with such small amounts. What effects there were resembled β-adrenergic action, unlike the actions of ethanol, and required quite high doses. This might have settled the matter, but the question has been reopened by the findings of Melchior and Myers, which will be described in Chapter 9. These compounds are still of considerable interest.

Summary

Acetaldehyde, the primary product of ethanol oxidation, circulates in the blood during ethanol metabolism. Although it is quite toxic, it does not contribute appreciably to the acute effects of ethanol. Its actions are sym-pathomimetic, mediated in part by release of norepinephrine. Disulfiram inhibits the enzymatic breakdown of acetaldehyde, thus allowing its accu-mulation after alcohol ingestion. The resultant dysphoric syndrome acts as a deterrent and is sometimes a successful component in the treatment of alcoholism. Acetaldehyde forms condensation products with biogenic amines; their pharmacological significance is doubtful at present.

References

1. Tottmar, S.O.C., Pettersson, H. and Kiessling, K.-H. The subcellular distri-bution and properties of aldehyde dehydrogenases in rat liver. Biochem. J. 135: 577–586, 1973.

2. Kalant, H., LeBlanc, A.E., Guttman, M. and Khanna, J.M. Metabolic and pharmacologic interaction of ethanol and metronidazole in the rat. Can. J. Physiol Pharmacol. 50: 476–484, 1972.

3. Sippel, H.W. and Eriksson, C.J.P. The acetaldehyde content in rat brain dur-ing ethanol oxidation. *In* The Role of Acetaldehyde in the Actions of Ethanol,

Lindros, K.O. and Eriksson, C.J.P., eds. Finnish Foundation for Alcohol Studies, Helsinki, 1975, pp. 149–157.

4. Goldberg, L. Quantitative studies on alcohol tolerance in man. Acta Physiol. Scand. 5: suppl. 16, 1943.

5. Beer, C.T. and Quastel, J.H. The effects of aliphatic aldehydes on the respiration of rat brain cortex slices and rat brain mitochondria. Can. J. Biochem. Physiol. 36: 531–541, 1958.

6. Eade, N.R. Mechanism of sympathomimetic action of aldehydes. J. Pharmacol. Exp. Ther. 127: 29–34, 1959.

7. James, T.N. and Bear, E.S. Effects of ethanol and acetaldehyde on the heart. Am. Heart J. 74: 243–255, 1967.

8. Egle, J.L., Jr. Effects of inhaled acetaldehyde and propionaldehyde on blood pressure and heart rate. Toxicol. Appl. Pharmacol. 23:131–135, 1972.

9. Hald, J., Jacobsen, E. and Larsen, V. The sensitizing effect of tetraethylthiuramdisulphide (Antabuse) to ethylalcohol. Acta Pharmacol. 4: 285–296, 1948.

10. Asmussen, E., Hald, J. and Larsen, V. The pharmacological action of acetaldehyde on the human organism. Acta Pharmacol. 4: 311–320, 1948.

11. Jacobsen, E. Is acetaldehyde an intermediary product in normal metabolism? Biochim. Biophys. Acta 4: 330–334, 1950.

12. Wilson, A., Davidson, W.J., Blanchard, R. and White, J. Disulfiram implantation. A placebo-controlled trial with two-year follow-up. J. Stud. Alc. 39: 809–819, 1978.

13. Davis, V.E. and Walsh, M.J. Alcohol, amines and alkaloids: a possible biochemical basis for alcohol addiction. Science 167: 1005–1007, 1970.

14. Goldstein, A. and Judson, B.A. Alcohol dependence and opiate dependence: lack of relationship in mice. Science 172: 290–292, 1971.

15. Cohen, G. Alkaloid products in the metabolism of alcohol and biogenic amines. Biochem. Pharmacol. 25: 1123–1128, 1976.

16. Collins, M.A. and Bigdeli, M.G. Tetrahydroisoquinolines in vivo: I. Rat brain formation of salsolinol, a condensation product of dopamine and acetaldehyde under certain conditions during ethanol intoxication. Life Sci. 16: 585–602, 1975.

17. Turner, A.J., Baker, K.M., Algeri, S., Frigerio, A. and Garattini, S. Tetrahydropapaveroline: formation in vivo and in vitro in rat brain. Life Sci. 14: 2247–2257, 1974.

18. O'Niell, P.J. and Rahwan, R.G. Absence of formation of brain salsolinol in ethanol-dependent mice. J. Pharmacol. Exp. Ther. 200: 306–313, 1977.

19. Hamilton, M.G., Blum, K. and Hirst, M. In vivo formation of isoquinoline alkaloids: effect of time and route of administration of ethanol. Adv. Exp. Med. Biol. 126: 73–86, 1980.

Review

Lindros, K.O. Acetaldehyde—its metabolism and role in the actions of alcohol. Res. Adv. Alc. Drug Prob. 4: 111–176, 1978.

3. The Liver

Chronic exposure to ethanol causes a variety of liver abnormalities, both functional and morphological, and cirrhosis of the liver, with its various complications, is an important cause of death in alcoholics. The relationship between cirrhosis and other, milder liver damage such as fatty liver or hepatitis is not yet entirely clear, but there is now a coherent body of knowledge about the biochemistry of alcoholic liver disease. This chapter will survey the main trends of experimental work exploring the mechanisms of fatty liver and alcohol hepatitis, as well as the beginnings of an understanding of cirrhosis.

Acute fatty liver

The fatty liver produced by single doses of ethanol is now quite well understood. It was discovered in the 1950s that rats accumulate fat in the liver after receiving a single huge dose of ethanol. Histologically, lipid droplets were visible within the hepatocytes, often progressing to the formation of a single large vesicle surrounded by a thin rim of cytoplasm and the displaced nucleus. Almost all the excess lipid was triglyceride; there was little change in phospholipids or cholesterol.

Mobilization of fatty acids from adipose tissue. Mallov and Bloch [1], who first showed this effect, examined the involvement of the adrenal and pituitary glands because they suspected a causative role of stress. Some of

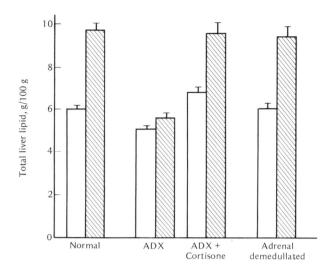

Fig. 3-1. *Acute fatty liver.* Female rats (which show this effect more strongly than males) were injected with a single dose of ethanol, 6.2 g/kg, and killed 17–19 hours later. Open bars represent rats given isocaloric amounts of glucose; stippled bars represent the ethanol-treated rats. Error bars are SEM for 8–39 rats. ADX = adrenalectomized. Note that total liver lipids are shown. If only the triglycerides are measured, more dramatic changes are seen (**Fig. 3-3**). From Mallov and Bloch [1].

their results are shown in Figure 3-1. At about 18 hours after a single large dose of ethanol, total liver lipids had increased substantially. Adrenalectomized rats did not show the acute fatty liver after alcohol administration. The adrenal cortex was the necessary component, since fatty liver after alcohol administration was seen in adrenalectomized rats treated with injections of cortisone. Hypophysectomy, like adrenalectomy, prevented fat accumulation, but in this case, cortisone was ineffective, thus indicating that some pituitary hormone other than adrenocorticotropic hormone (ACTH) must be necessary.

A few years later, B.B. Brodie and his colleagues at the National Institutes of Health confirmed and extended these findings with a series of experiments in support of the hypothesis that a stress-induced mobilization of fatty acids from adipose tissue causes the fatty liver [2]. To understand this, it is useful to consider the pathways by which fatty acids enter and leave the free fatty acid and triglyceride pools of the liver and adipose tissue (Fig. 3-2). The depot fats are replenished by triglycerides stripped

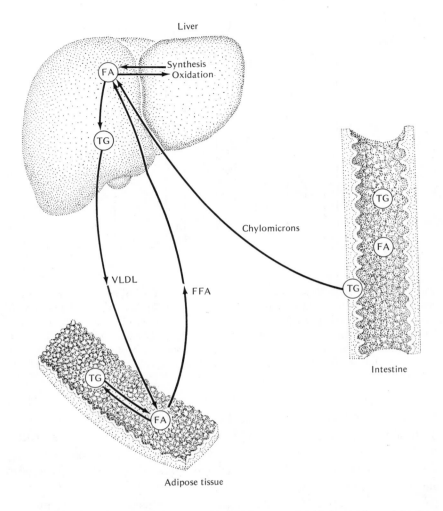

Fig. 3-2. *Major pathways of fatty acid metabolism.* Dietary triglycerides are broken down in the intestinal wall and immediately reassembled for transport in chylomicrons. Reaching the liver, they are again hydrolyzed and enter the hepatic pool of free fatty acids. The same pool is also fed by free fatty acids (actually bound to plasma albumin) arriving from the adipose tissue after stored triglycerides have been hydrolyzed by tissue lipases and by synthesis of fatty acids in the liver, a process requiring NADH. Fatty acids leave the hepatic pool by being oxidized, an NAD-dependent step, or by being esterified into triglycerides again. These fats are exported from the liver in the form of very low density lipoproteins, VLDL, and return to adipose tissue.

from plasma lipoproteins and simultaneously they are broken down by lipases, releasing free fatty acids that serve as important sources of energy for the heart and liver. When the supply of free fatty acids exceeds the liver's immediate needs, they are reassembled into triglycerides for export in the form of lipoproteins, and they return to the adipose tissue for storage. This continual recycling process fuels the body between meals. In times of stress, it is accelerated. Brodie showed that the acute fatty liver was part of a general stress effect that is mediated by the sympathetic nervous system. In rats, a rise in plasma free fatty acids and in liver triglycerides followed a single large dose of ethanol. The same thing happened after a big dose of morphine or a nonspecific stressful experience, such as exposure to cold. Adipose tissue was the source of the fatty acids that accumulated as triglycerides in the liver. Injection of epinephrine or norepinephrine, or stimulation of the nerves leading to adipose tissue could cause release of free fatty acids. A ganglionic blocking agent or an α-adrenergic blocker prevented the rise in plasma free fatty acids as well as the fatty liver after acute ethanol treatment. Finally, it was shown that norepinephrine activates a lipase in adipose tissue, releasing free fatty acids from triglycerides. The lipase activity, measured *in vitro*, was elevated in tissue taken from rats that had been treated with ethanol (or morphine or cold). *In vitro*, the lipase itself was stimulated by catecholamines.

Hepatic metabolism of fatty acids. While Brodie's group was working out the details of the oversupply of fatty acids to the liver, Isselbacher and co-workers were studying the handling of fatty acids in the liver [3]. Fatty acids taken up from the plasma and those newly synthesized in the liver enter a common pool, which they leave when they are oxidized or esterified. Normally, steady state conditions prevail, and the fatty acid and triglyceride pools are of constant size. Isselbacher and his colleagues showed that ethanol affects several components of the steady state. The input of fatty acids into the liver pool is increased as shown above. Synthesis and oxidation of free fatty acids are affected in opposite directions by the shift in the redox balance during metabolism of ethanol (Chapter 1). An increased rate of fatty acid synthesis from radioactive acetate was demonstrated after *in vivo* administration of ethanol and was ascribed to the increased availability of NADH. To study the oxidation of fatty acids, these workers measured the conversion of labeled palmitate to carbon dioxide *in vivo* and *in vitro*. The oxidation of palmitate was partially blocked by ethanol, presumably because of the relative lack of NAD during the metabolism of ethanol. Addition of NAD to the homogenate restored palmitate oxidation *in vitro* in the presence of ethanol. Esterification of fatty acids was stimulated by ethanol *in vivo* but not *in vitro*. Finally, the release of triglycerides from the isolated, perfused rat liver was also found to be affected [4]. When rat livers were perfused with labeled palmitate, some of the label appeared in the perfusate in the form of triglycerides packaged in lipoproteins. Ethanol added to the infusion fluid blocked the removal of triglycerides from the liver. Thus it appears that many different steps in the handling of fatty acids are affected by acute ethanol. The liver, stoked with extra fatty acids to take care of the stressful emergency, is preoccupied with oxidation of ethanol and cannot handle the influx. Triglycerides accumulate rapidly.

A weekend's heavy drinking is probably enough to induce a fatty liver in humans. Rubin and Lieber [5] administered ethanol for two days to non-alcoholic human volunteers, along with a more than adequate diet (low fat and high protein). Liver triglycerides doubled, as shown by biopsies before and after alcohol administration. Blood alcohol levels remained rather low (0.2–0.8 mg/ml), and it was pointed out that one need not even be drunk to damage one's liver.

Chronic effects of ethanol

Chronic ethanol consumption can produce a variety of morphological abnormalities, including fatty liver, hepatitis, necrosis, and fibrosis, along with impaired liver function. The sequence and causality of these various conditions in alcoholic patients are unclear. In particular, the role of malnutrition is difficult to assess in the population of alcoholics admitted to the hospital. It is well known that malnutrition can cause most of the types of liver damage seen in alcoholics and also that alcoholics are often malnourished because they can obtain most of their calories from the ethanol they drink. At 7.1 kilocalories per gram of ethanol, a pint of whiskey supplies about 1300 kilocalories, a substantial part of a daily energy supply, but unaccompanied by other essential components of a good diet. Alcoholics enter the hospital with nutritional deficiencies of protein, choline, vitamins, etc., that are otherwise rare in this country today. But not all alcoholics are undernourished. Clearly we need animal models to sort out the variables.

A basic difficulty in any animal experiment where ethanol is administered chronically is that effective doses of ethanol necessarily displace a considerable proportion of food calories from the diet. If ethanol is given by injection (or intubation) or is included in the animals' drinking water, the treated animals will eat less solid food than the untreated controls. Rats, unlike humans, regulate their caloric intake within narrow limits. Proper controls should receive the same amounts and proportions of fat, protein, and carbohydrate in their diet as the alcohol-treated animals. The introduction of totally liquid diets by Lieber and co-workers [6] was a big advance. The liquid diet contained optimal amounts of fat and protein, to which either ethanol or another carbohydrate was added in isocaloric quantities. The alcohol-treated animals and their controls then had exactly the same balance among the components of the diet. By pair-feeding, the animals of the two groups can be made to take exactly the same amounts as well as the same proportions of nutrients. Pair-feeding means that each treated animal has a partner whose daily supply of the control diet is limited to the amount of the alcohol diet taken by its mate in the experimental group. The control group runs a day behind the alcohol group, sometimes for months, receiving each day the amount of diet taken by the treated animals the day before.

Chronic fatty liver. Lieber and colleagues studied the chronic fatty liver in rats treated for 24 days with a diet containing 36% of the calories as ethanol [7]. Liver triglycerides increased, even when the diet contained almost no fat. When the diet contained a fairly high amount of fat (43% of the calories), fatty liver was more pronounced. An important finding was that the source of the fatty acids in this chronic fatty liver was no longer the adipose tissue as had been the case in the acute fatty liver. Instead, the fatty acids came from the diet and from endogenously synthesized fatty acids. This was shown in an ingenious experiment where the fatty acid composition of the adipose stores differed from that of the diet supplied during alcohol administration. The adipose tissue was prelabeled by feeding the animals either coconut oil, which is rich in short-chain saturated acids such as lauric (12:0) and myristic (14:0), or linseed oil, which contains 50% linolenic acid (18:3). After three weeks, these distinctive fatty acids constituted 30% of the depot fats. Then the diets were switched and alcohol was included in the diet of half of each group for an 11-day alcohol administration period. Figure 3-3 shows that the liver now contained fatty acids typical of the new diet, quite a different composition from that of the adipose tissue. These experiments were important in demonstrating that ethanol itself was hep-

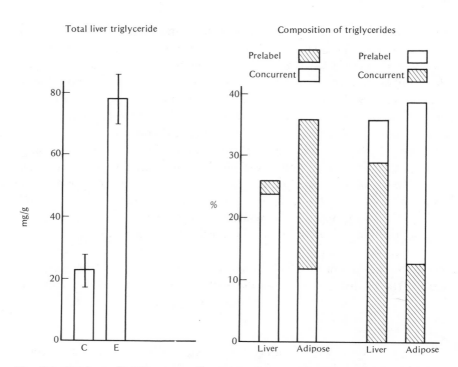

Fig. 3-3. *Triglycerides that accumulate in the liver after chronic ethanol treatment resemble the diet, not the adipose tissue.* Rats were fed ethanol in a liquid diet for 11 days; the resulting tripling of the triglyceride content of the liver is shown at the left, in comparison to controls pair-fed a sucrose diet. The four bars at the right show results of switching the fatty acid composition of the diet at the time of addition of ethanol. When the adipose tissue was prelabeled with a diet rich in 12:0 and 14:0 (stippled bar) and then the diet was changed to one high in linolenate (blank bar) during the alcohol administration phase, the liver accumulated linolenate from the current diet, not the short-chain fatty acids stored in the adipose tissue. The bars on the far right show the converse experiment. Again the ethanol-exposed liver gained fatty acids from the diet (or endogenous synthesis), not from the prelabeled adipose tissue. From Lieber *et al.* [7].

atotoxic, even in the presence of an adequate diet. The same conclusion was reached when fatty livers were produced in human alcoholic subjects to whom ethanol was administered for two or three weeks on the ward [8].

Alcoholic hepatitis. Hepatitis is an inflammatory condition, i.e., a response to injury, characterized primarily by infiltration of polymorphonuclear leukocytes. Alcoholic hepatitis is recognized clinically by anorexia, fever, jaundice, and pain. The liver may be enlarged and its function impaired, as shown by serum enzyme levels. Histologically, hepatitis is easily recognized. Along with steatosis, clusters of polymorphonuclear leukocytes are observed, especially near the central vein, and at the same region there may be areas of necrosis. A characteristic feature, though not specific for alcohol damage, is the hyalin inclusion or Mallory body. This is a glassy material of irregular outline, inside the hepatocytes; its origin and fate are unknown. With the electron microscope, branched microfilaments are seen which do not seem to be actin or tubulin, but which might be derived from other normal filamentous proteins [9].

Liver mitochondria in patients with alcoholic hepatitis are characteristically abnormal, often greatly enlarged and sometimes distorted in shape; the cristae are shortened and in disarray. A quantitative electron micrograph study [10] in rats that had been chronically treated with ethanol showed that the mean cross-sectional area of mitochondria was increased from 0.40 to 0.58 square microns, which corresponds to almost a two-fold increase in volume. The cristae were shortened by half. There was no change in the number of mitochondria.

Electron micrography also shows changes in endoplasmic reticulum in alcoholic hepatitis. The rough endoplasmic reticulum is disrupted and decreased in amount, whereas the smooth endoplasmic reticulum is greatly hypertrophied. The increased smooth endoplasmic reticulum corresponds to the elevation of mixed function oxygenase activity that accompanies induction of MEOS, since microsomal drug-metabolizing enzymes occupy the smooth endoplasmic reticulum fraction.

Animal models of alcohol hepatitis have not been easy to develop. No one has produced a convincing alcoholic hepatitis in rats except when pyrazole was administered along with the ethanol [11]. Rats treated with ethanol alone show a fatty liver after a couple of weeks on the liquid diet, but no serious pathology. Addition of a small dose of pyrazole every other day produced more abnormalities, such as focal necrosis, infiltration with polymorphonuclear leukocytes, and hyalin inclusions. A few distorted mitochondria were seen. The rough endoplasmic reticulum was vesiculated and not in its usual orderly arrangement, and there was conspicuous hypertrophy of the smooth endoplasmic reticulum. Centrilobular regions were specifically vulnerable to these changes. These animals were kept on ethanol diets for 11 months. Surprisingly, the alcoholic hepatitis regressed after two months, despite continued treatment with ethanol and pyrazole. Hydroxyproline, an index of collagen content and thus a sensitive measure of incipient fibrosis, was not increased by the alcohol treatment. The liver pathology occurred in the presence of pyrazole, indicating that it was caused by ethanol itself rather than by its direct or indirect metabolic products. Pyrazole alone caused no damage.

Meanwhile, Lieber's group was administering ethanol to baboons, a species presumed to handle fats as humans do. Three baboons were treated with a liquid diet containing 50% of calories as ethanol for nine months [12]. Controls were pair-fed on a sucrose diet, isocaloric with the ethanol. At the end of this time, serum glutamic-oxalacetic transaminase (SGOT) levels indicated impairment of liver function in the ethanol-treated animals. Further, biopsies of the liver showed evidence of alcoholic hepatitis, including fatty accumulation, inflammation, hyaline, and necrotic zones around the central veins. There was also some fibrosis and it seemed that cirrhosis could not be far behind.

By now, it was clear that ethanol was indeed directly toxic to the liver. Hartroft, Porta, and others had been arguing to the contrary, presenting experiments to show that the effects of ethanol could be overcome with good nutrition. It now seems clear that either ethanol or malnutrition can damage the liver and that many alcoholic patients have damage from both causes. The comforting notion that the effects of heavy drinking can be offset by heavy eating is no longer tenable.

Hypermetabolic state. We turn now to a new approach that offers a possibility of understanding the necrotic process in biochemical terms, and even some hope of developing rational preventive or therapeutic measures. Israel and co-workers have described a "hypermetabolic state" of the liver after chronic treatment with ethanol. In a series of investigations since 1973, they have developed the hypothesis that the liver of a rat chronically treated with ethanol is in a functionally hyperthyroid state. Livers of rats treated chronically with ethanol had increased activity of Na,K-activated ATPase and an increased rate of oxygen uptake [13]. The extra oxygen uptake was ouabain sensitive, indicating that it was caused by the overactive ATPase. The phosphorylation potential, $[ATP]/([ADP] \times [Pi])$, was decreased, a condition that stimulates respiration. There was a striking biochemical similarity between such livers and those of rats given a single dose of thyroxine, 24–30 hours before sacrifice [14]. Like ethanol, thyroxine increased the activity of hepatic Na,K-ATPase and the rate of oxygen uptake by liver slices, and ouabain blocked these effects. Oxidation of ethanol itself by liver slices was increased after the chronic ethanol administration, but a comparable effect of thyroxine was seen only at low doses, since high levels of thyroxine inhibit liver alcohol dehydrogenase.

The hypermetabolic state could be produced by other conditions as well [15]. Exposure to cold produced it, and so did single doses of epinephrine, an effect mediated by α-receptors. It did not occur in adrenalectomized animals, but we do not yet know whether it is the adrenal medulla or cortex that is required. While it is now clear that the liver of alcohol-treated rats closely resembles a hyperthyroid one, it is not yet known whether the ethanol effect is mediated by thyroid hormones or whether ethanol mimics the action of the thyroid, perhaps acting directly on the ATPase.

On the basis of these findings, a mechanism for ethanol-induced liver toxicity has been proposed and tested. In the hypermetabolic state, an additional strain might be laid on the liver by hypoxia. Such damage should occur most readily in the regions of the liver with the lowest oxygen supply, i.e., the centrilobular area near the hepatic vein and farthest from the incoming blood supply of the portal vein and hepatic artery. There is a gradient of oxygen between the portal vein–hepatic artery input and the hepatic vein output (Fig. 3-4). Experimentally, exposure to low oxygen tension for a few hours had no effect on the liver of normal rats, but the livers of rats that had been chronically treated with ethanol showed centrilobular necrosis as in alcoholic hepatitis, along with a significant increase in SGOT. Thus, an episode of hypoxia, such as might result from pulmonary infection, respiratory depression, etc., may be the last straw for a liver that is using up oxygen very fast. Necrosis appears at the region of lowest oxygen tension in this and several other forms of liver injury.

The possibility that necrosis might be linked to the liver hypermetabolic state suggested that antithyroid drugs might counteract the damage. Israel et al. [17] found that propylthiouracil abolished the hypermetabolic state and the functional impairment, and greatly reduced the necrosis when alcohol-treated animals were subjected to low oxygen tension. This encouraging finding indicates a real possibility of new drug treatment for liver damage. Propylthiouracil acts partly by inhibiting the synthesis of thyroxine in the thyroid gland but mainly by blocking the conversion of thyroxine (T_4) to the much more active triiodothyronine (T_3), a process that occurs in the liver and kidneys.

This is still a controversial area of research and we must wait for confirmation from other laboratories before feeling confident about it. Some investigators do not see the hypermetabolic state in isolated hepatocytes or

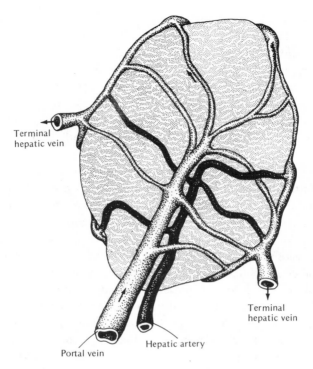

Terminal
hepatic vein

Terminal
hepatic vein

Hepatic artery

Portal vein

Fig. 3-4. *Liver microcirculation.* The functional unit of the liver is the acinus, one of which (grossly oversimplified) is diagrammed here. Blood enters an acinus through the portal vein and the hepatic artery, which travel together along with the terminal bile ducts (not shown). The blood courses through sinusoids among the liver parenchymal cells to join the terminal hepatic veins at the periphery of the acinus. The oxygen content of the blood is highest near the portal vein and hepatic artery, and lowest near the hepatic vein. In older terminology, the liver lobule was seen as a unit surrounding the "central" vein, which is here called the terminal hepatic vein. What matters is that the oxygenation is least near the vessel that drains blood from the liver and that this is also the site of necrosis in alcoholic liver damage. Adapted from Rappaport [16].

in perfused livers from rats that have been chronically treated with ethanol, [18,19].

Double-blind clinical trials of propylthiouracil showed improvement in liver function and clinical condition in patients with alcoholic hepatitis [20]. Patients with the greatest degree of initial impairment benefited most, but those with cirrhosis were unaffected. It turned out that these patients did not have excess circulating thyroid hormone after all. On the contrary, their serum T_3 levels were low on admission, more so in the more severely ill patients, as Figure 3-5 shows, reflecting impairment of T_3 synthesis by the liver [21]. T_4 levels were generally unaffected. Nevertheless, propylthiouracil therapy was beneficial. This exciting story is clearly incomplete.

Cirrhosis. Cirrhosis is the last and deadliest of the liver injuries that ethanol can cause. The cirrhotic liver is apparently the result of a long period of intermittent liver injury, causing focal necrosis and scar formation, while other areas simultaneously regenerate. Such a liver is small, hard, and lumpy, in contrast to the large, smooth, fatty liver of the earlier stages. Histologically, it is easy to see the extensive fibrosis that obliterates the regular structure of the lobules or acini. Fibrosis may obstruct the portal

Fig. 3-5. *Decreased T_3 in serum of patients with liver disease.* The degree of clinical impairment in patients with a diagnosis of alcoholic liver disease was estimated by a rating scale known as the composite clinical and laboratory index (CCLI), which included clinical signs such as hepatomegaly and encephalopathy, plus laboratory findings such as SGOT and prothrombin time. Plotting admission values of serum T_3 against the CCLI gave a highly significant negative correlation, suggesting that impaired livers fail to convert T_4 to T_3. From Israel et al. [19].

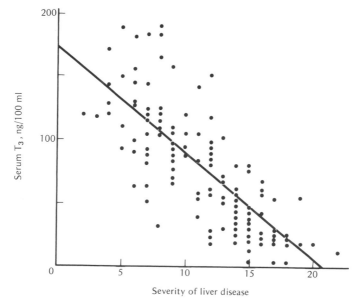

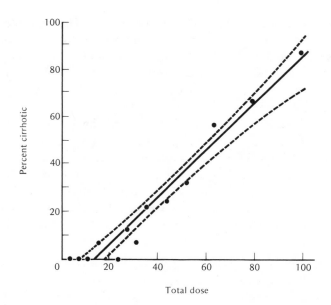

Fig. 3-6. *Dose-relation of liver cirrhosis.* Drinking histories were obtained from 265 alcoholics. The total amount of ethanol consumed during each individual's drinking life was expressed as the product of mean daily intake (g/kg) and years of drinking. Individuals were assigned to one of 13 dose-intervals and the incidence of cirrhosis in each class was computed. Although only 39 cases of cirrhosis were found, an excellent correlation was observed. The least-squares line and its confidence limits are shown. From Lelbach [21].

vein, leading to portal hypertension, with consequent ascites and varicose veins in the gastrointestinal tract. The varices may rupture, with fatal results. The cirrhotic liver does not function well, and its ultimate failure is a frequent cause of death in alcoholics. By no means do all alcoholics have cirrhotic livers. It has been hard to explain why only a small percent of longtime heavy drinkers develop this condition. Something accidental may trigger it, such as an episode of hypoxia, in the mechanism just described. Lelbach has been able to relate cirrhosis rates to histories of alcohol intake [22]. The incidence of cirrhosis increases progressively among groups of patients with increasing exposure to ethanol, both in terms of duration and intensity of drinking. The type of beverage does not seem to matter, but the data suggest that continuous drinking may be more dangerous than periodic drinking.

Rubin and Lieber have finally achieved the production of cirrhosis itself in an animal model [23]. They did not succeed with rats, despite a vigorous attempt. Livers of rats kept on an alcohol liquid diet for 14 months had increased hydroxyproline content and evidence of accelerated collagen synthesis, but no visible fibrosis. Baboons were maintained on alcohol for up to four years. The hydroxyproline content of their livers increased faster than in rats; this may indicate that the primate is indeed a more suitable model. All 13 of the alcohol-treated baboons had fatty livers and most had inflammatory changes, including necrosis and elevated SGOT values. One animal clearly had a cirrhotic liver and another had an "incomplete" cirrhosis.

Summary

The fatty liver produced by single large doses of ethanol is simply a stress reaction. The adrenals mediate a mobilization of fatty acids from adipose tissue while the liver is preoccupied with ethanol oxidation and cannot handle the influx of fatty acids. Chronic alcoholic fatty liver is an effect of ethanol per se, although similar pathology can also be caused by malnutrition. It may be followed by alcoholic hepatitis, with inflammation and centrilobular necrosis. These changes can be reproduced in animal models, most successfully in baboons. A hypermetabolic state of the liver after chronic treatment with ethanol has been described. It resembles a hyperthyroid state, although it has been shown that patients with alcoholic disease actually have low circulating thyroid hormones. The hypermetabolic state may be due to induction of liver Na,K-ATPase, and it may be the cause of necrosis in areas of the liver lobule where hypoxia may occur adventitiously. Experimental treatment with an antithyroid drug seems to be beneficial to patients with alcoholic hepatitis.

Many questions remain to be answered about alcoholic liver disease, but the progress is encouraging, both in the development of animal models and in understanding biochemical mechanisms. The liver is complex, but less so than the brain, and we do have some hope of understanding the clinically important hepatotoxicity of ethanol.

References

1. Mallov, S. and Bloch, J.L. Role of hypophysis and adrenals in fatty infiltration of liver resulting from acute ethanol intoxication. Am. J. Physiol. 184: 29–34, 1956.

2. Brodie, B.B. and Maickel, R.P. Role of the sympathetic nervous system in drug-induced fatty liver. Ann. New York Acad. Sci. 104: 1049–1058, 1963.

3. Reboucas, G. and Isselbacher, K.J. Studies on the pathogenesis of the ethanol-induced fatty liver. I. Synthesis and oxidation of fatty acids by the liver. J. Clin. Invest. 40: 1355–1362, 1961.

4. Schapiro, R.H., Drummey, G.D., Shimizu, Y. and Isselbacher, K.J. Studies on the pathogenesis of the ethanol-induced fatty liver. II. Effect of ethanol on palmitate-1-C^{14} metabolism by the isolated perfused rat liver. J. Clin. Invest. 43: 1338–1347, 1964.

5. Rubin, E. and Lieber, C.S. Alcohol-induced hepatic injury in nonalcoholic volunteers. New Eng. J. Med. 278: 869–876, 1968.

6. DeCarli, L.M. and Lieber, C.S. Fatty liver in the rat after prolonged intake of ethanol with a nutritionally adequate new liquid diet. J. Nutrition 91: 331–336, 1967.

7. Lieber, C.S., Spritz, N. and DeCarli, L.M. Role of dietary, adipose, and endogenously synthesized fatty acids in the pathogenesis of the alcoholic fatty liver. J. Clin. Invest. 45: 51–62, 1966.

8. Lieber, C.S., Jones, D.P. and DeCarli, L.M. Effects of prolonged ethanol intake: production of fatty liver despite adequate diets. J. Clin. Invest. 44: 1009–1021, 1965.

9. French, S.W. and Davies, P.L. The Mallory body in the pathogenesis of alcoholic liver disease. In Alcoholic Liver Pathology, Khanna, J.M., Israel, Y. and Kalant, H., Eds. Addiction Research Foundation, Toronto, 1975, pp. 113–144.

10. Kiessling, K.-H. and Pilström, L. Effect of ethanol on rat liver. II. Number, size and appearance of mitochondria. Acta Pharmacol. Toxicol. 24: 103–111, 1966.

11. Phillips, M.J., Chiu, H.F., Khanna, J.M. and Kalant, H. Chronic effects of ethanol pyrazole administration on rats fed a nutritionally adequate diet. In Alcoholic Liver Pathology, Khanna, J.M., Israel, Y. and Kalant, H., Eds. Addiction Research Foundation, Toronto, 1975, pp. 271–288.

12. Rubin, E. and Lieber, C.S. Experimental alcoholic hepatitis: a new primate model. Science 182: 712–713, 1973.

13. Bernstein, J., Videla, L. and Israel, Y. Metabolic alterations produced in the liver by chronic ethanol administration. Changes related to energetic parameters of the cell. Biochem. J. 134: 515–521, 1973.

14. Israel, Y., Videla, L., MacDonald, A. and Bernstein, J. Metabolic alterations produced in the liver by chronic ethanol administration. Comparison between the effects produced by ethanol and by thyroid hormones. Biochem. J. 134: 523–529, 1973.

15. Bernstein, J., Videla, L. and Israel, Y. Hormonal influences in the development of the hypermetabolic state of the liver produced by chronic administration of ethanol. J. Pharmacol. Exp. Ther. 192: 583–591, 1975.

16. Rappaport, A.M. The microcirculatory acinar concept of normal and pathological hepatic structure. Beitr. Path. 157: 215–243, 1976.

17. Israel, Y., Kalant, H., Orrego, H., Khanna, J.M., Videla, L. and Phillips, J.M.

Experimental alcohol-induced hepatic necrosis: suppression by propylthioura-cil. Proc. Nat. Acad. Sci. 72: 1137–1141, 1975.

18. Cederbaum, A.I., Dicker, E., Lieber, C.S. and Rubin, E. Ethanol oxidation by isolated hepatocytes from ethanol-treated and control rats: factors contributing to the metabolic adaptation after chronic ethanol consumption. Biochem. Pharmacol. 27: 7–15, 1978.

19. Schaffer, W.T., Denckla, W.D. and Veech, R.L. Effects of chronic ethanol administration on O_2 consumption in whole body and perfused liver of the rat. Alcoholism Clin. Exp. Res. 5: 192–197, 1981.

20. Orrego, H., Kalant, H., Israel, Y., Blake, J., Medline, A., Rankin, J.G., Armstrong, A. and Kapur, B. Effect of short-term therapy with propylthiouracil in patients with alcoholic liver disease. Gastroenterology 76: 105–115, 1979.

21. Israel, Y., Walfish, P.G., Orrego, H., Blake, J. and Kalant, H. Thyroid hormones in alcoholic liver disease. Effect of treatment with 6-n-propylthiouracil. Gastroenterology 76: 116–122, 1979.

22. Lelbach, W.K. Quantitative aspects of drinking in alcoholic liver cirrhosis. In Alcoholic Liver Pathology, Khanna, J.M., Israel, Y. and Kalant, H., eds. Addiction Research Foundation, Toronto, 1975, pp. 1–18.

23. Lelbach, W.K. Dosis-Wirkungs-Beziehung bei Alkohol-Leberschäden. Dtsch. med. Wschr. 97: 1435–1436, 1972.

24. Rubin, E. and Lieber, C.S. Fatty liver, alcoholic hepatitis and cirrhosis produced by alcohol in primates. New Eng. J. Med. 290: 128–135, 1974.

Review

Feinman, L. and Lieber, C.S. Liver disease in alcoholism. In The Biology of Alcoholism, Kissin, B. and Begleiter, H., eds., Vol. 3. Plenum Press, New York, 1974, pp. 303–338.

4. Biophysical Pharmacology: Alcohol Effects on Biomembranes

We will now consider the search for the actual molecular mechanism of action of this chemically nondescript drug. This brings us into a new field, biophysical pharmacology. Ethanol belongs to a class of drugs that act more like physical agents than like chemicals. Other than alcohol dehydrogenase, we know of no macromolecule that can react with ethanol with any degree of specificity, and it seems unlikely that a chemical receptor mediates the actions of this little drug. In studying ethanol, we use many kinds of pharmacology, but receptor pharmacology, the study of specific interactions of small molecules with big ones, is hardly to be found here. It is only a small part of the ethanol story, though for many other drugs it is the approach that has been most fruitful in the past few decades. In this chapter, we will consider drug effects that do not depend on specificity of chemical structure.

The Meyer-Overton concept

The first membrane pharmacology was done decades ago. A basic observation made independently by E. Overton and H.H. Meyer in about 1900 was that the potencies of anesthetic drugs are directly proportional to their lipid solubilities [1]. They measured potency in terms of the effective aqueous concentrations in crude biological systems, such as overturning tadpoles. (Aquatic animals can be useful in pharmacology. Drug concentrations in the animal can be held constant without regard to elimination rates if, instead of injecting the drug into the animal, one puts the animal into the drug.) Lipid solubility was measured as the olive oil:water partition coefficient, the ratio of the drug concentrations in lipid and aqueous phases. The proportionality between effective concentrations and lipid solubility applied to a large group of drugs, including alcohols, whose potency range extended over four or five orders of magnitude. The drugs that Overton and Meyer used were a variety of simple organic compounds, not clinically used anesthetics. They interpreted their findings to mean that these drugs must act in some lipid portion of the body. "Narcosis commences when any chemically indifferent substance has attained a certain molar concentration in the lipoids of the cell." It was later surmised that this site must be the cell membrane, but it was not clear at first whether the drugs must have good solubility in order to penetrate through membranes (and thus get into the brain, for example), or whether their action was within the membranes. The Meyer-Overton concept was an idea so far ahead of its time that little could be done with it for decades. Only recently have the biophysical methods been developed that allow us to study what drugs might do in the lipid environment. Meanwhile, however, some refinements were made in the Meyer-Overton hypothesis. Mullins pointed out that even better correlations with potency could be made if the molecular volumes were taken into account [2]. That is, it is not only the molar concentration of the drug in the membrane that counts, but also the amount of space it occupies. Mole for mole, bigger molecules are more effective. This tells us that the drugs are indeed acting within the membrane and not just passing through. Drugs of this class are a motley group indeed. Many have simple structures like ethanol; otherwise they do not necessarily resemble each other. They include general anesthetics, such as halogenated hydrocarbons, alcohols, and nitrous oxide.

It was Seeman who developed membrane pharmacology as a modern discipline [3]. In the late 1960s, he studied the ability of drugs to expand erythrocyte membranes, and he found that many anesthetic drugs, including several alcohols, would protect red blood cells against hemolysis in hypotonic solutions [4]. This antihemolytic effect was explained by expansion of the membrane area, so that the cell could swell without bursting. The critical lytic volume was shown to be enlarged by the drugs, as illustrated in Figure 4-1. The membrane area (calculated from the expansion of critical lytic cell volume) was increased by about 3% at the nerve-blocking concentration. In this system, drug potency correlated well with lipid solubility and even better with the volume of drug occupying the membrane.

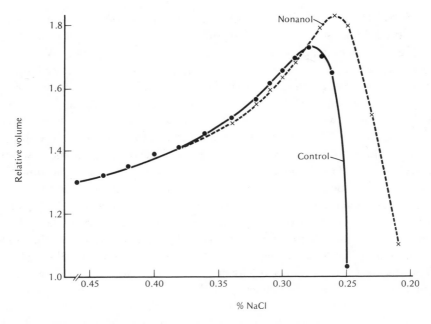

Fig. 4-1. *Alcohols expand the critical lytic volume of erythrocytes.* At various hypotonic concentrations of NaC1, the mean cell volume was determined from careful measurements of cell counts and hematocrits. As the tonicity was lowered, the cells swelled and finally lysed, whereupon the apparent volume fell sharply. The volume at which lysis began was increased by 6% by the presence of 0.34 mM nonanol. The alcohol did not affect the osmotic swelling but protected the cells against hemolysis by expanding the area of the cell membrane. From Seeman et al. [4].

Ethanol lines up nicely with other drugs in Meyer-Overton plots (Fig. 4-2). This is our first indication that ethanol itself acts in membranes. Seeman later did a direct experiment with ethanol, using a high-precision densitometer to show that ethanol lowers the density of erythrocyte, synaptosomal, and model membranes [5]. In the pure cholesterol/lecithin membranes, the expansion of volume could be accounted for by the amount of ethanol occupying space in the membrane, but in biological membranes ethanol had a much greater effect, suggesting that conformational changes in proteins contributed to the swelling.

The Singer-Nicolson model

A basis for understanding the expansion of cell membranes by alcohol is the Singer-Nicolson model of a generalized cell membrane, the fluid mosaic model [6]. This is a phospholipid bilayer containing cholesterol and proteins (Fig. 4-3). The polar phospholipids have charged, hydrophilic head groups that stay in the aqueous phase, at the interface of the membrane with the cytosol or the extracellular fluid. The long acyl chains of the phospholipid fatty acids are strongly hydrophobic. Seeking each other rather than the aqueous phase, they gather at the center of the bilayer. Scattered in the membrane, in different concentrations for different cell membrane types is cholesterol, which has a stiff cylindrical steroid nucleus with a hydroxyl group at one end, near the membrane surface. Its acyl chain extends inward, along with those of the phospholipids. Some cells (erythrocytes, for example) have nearly as many molecules of cholesterol as phospholipid

Fig. 4-2. A *Meyer-Overton plot*. This scatter plot of lipid solubility (expressed as membrane-buffer partition coefficient) against potency shows an excellent correlation. Potency is here represented by its reciprocal, the concentration necessary to produce a certain effect, either nerve block or a specified degree of antihemolytic action. Seeman had shown that the two properties go together. Over a wide range of aqueous concentrations, the most potent drugs are those with the highest lipid solubility. The normal alcohols are indicated here by their chain lengths, C1 to C10. This shows that ethanol is one of the drugs that fit the Meyer-Overton concept. From Seeman [3].

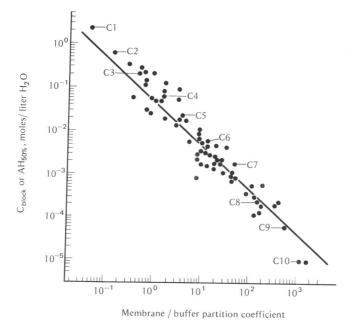

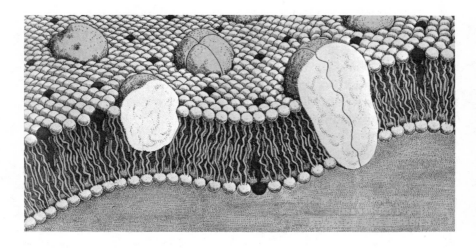

Fig. 4-3. *Components of a typical membrane.* The diagram shows a phospholipid-cholesterol bilayer matrix in which proteins are embedded to various depths. Light colored circles indicate phospholipid head groups and dark circles represent cholesterol. Acyl chains extend into the center of the bilayer.

in their plasma membranes. Many other lipids exist in cell membranes; their structural and functional roles have not been discovered, and their pharmacological significance is unknown. Proteins, embedded to various depths in the lipid sheet, make up more than half the membrane by weight. Peripheral proteins are loosely attached to the membrane surface, while integral proteins are deeply sunk into the hydrophobic region and often penetrate the bilayer completely. Some integral proteins may be surrounded by a shell of "boundary lipid" that is less fluid than the rest of the lipid matrix [7].

Membrane proteins carry out various kinds of transport. Some of the proteins are channels or ionophores through which ions can move by passive diffusion. These may be proteins that have (in their folded form) a hydrophobic exterior and a hydrophilic core, lined with fixed charges, which actually forms a hole through which small ions can penetrate. Other proteins are carriers, or energy-requiring pumps. The best-known of these is the Mg-dependent Na,K-activated ATPase that occurs in most mammalian cell membranes. It pumps sodium out and potassium in, to maintain the normal gradient between the cytosol and the extracellular fluid. This is particularly important in nerves, to restore the membrane potential after a series of action potentials. Coupled protein systems exist in brain membranes. For example, the β-adrenergic receptor on the other monolayer is functionally linked to adenylate cyclase whose catalytic unit may be in the inner monolayer. The GABA binding site, the separate binding sites for benzodiazepines and some convulsant drugs, and the chloride ionophore seem to be parts of a single functional unit, held together perhaps by its lipid environment. Another kind of membrane transport is endo- or exocytosis, whereby large particles are taken through membranes, e.g., for release of neurotransmitters by fusion of storage vesicles with the cell membrane. Protein-mediated transport functions are the direct or indirect targets of drugs that act on membranes.

Effects of ethanol and related drugs on membrane fluidity

The Singer-Nicolson model lets us easily visualize the physical properties of membranes that might be affected by drugs. The cell membrane is by no means a rigid shell. The lipids are in a semiliquid state and considerable jostling about is possible in the plane of the membrane [8]. A molecule can change places with its neighbor in the bilayer very rapidly. This lateral diffusion is clearly important for function. Membrane components must be free to combine with or depart from one another, for allosteric transformations of membrane enzymes by association of subunits and for coupling of components located in the two halves of the bilayer. Lateral diffusion can be observed by the spread of a spot of spin label in a membrane or by the migration of labeled antibody in a hybrid mouse-human cell. Clearly, the degree of fluidity is important for lateral diffusion. A stiff gel membrane would not allow it, and too fluid a membrane might be chaotic. In contrast to the speed of lateral diffusion, motion through the membrane from one surface to the other (flipflop) is much more restricted. It happens extremely slowly, if at all, in model membranes, but may occur rapidly in living cells, by processes that are not yet well understood. Energy is necessary to pull charged groups through the hydrophobic interior of a membrane, and this would require some special mechanism.

Membrane disordering by drugs. The general fluidity of membranes can be studied with nuclear magnetic resonance, electron paramagnetic resonance, and fluorescence polarization. All three methods have been used to show that alcohols disorder membrane lipids.

The first application of modern instrumentation to membrane pharmacology was a nuclear magnetic resonance study of benzyl alcohol in red cell membranes [9]. Using the proton signal from the benzyl alcohol itself, it was found that the drug caused the membrane to become more disordered. The disordering effect could also be seen in membranes made of the lipids extracted from the erythrocytes. Thus, an interaction of the drug with the membrane lipids caused the disordering.

Electron paramagnetic resonance techniques (EPR, also called electron spin resonance) depend on absorption of energy by a spin label probe (unpaired electron) in a magnetic field. The spin label is attached to a compound that can be taken up into a membrane. There it reports its own motion. For example, a spin label derivative of stearic acid can be incorporated into membranes; from its EPR spectrum one can calculate an order parameter, a quantity that indicates the degree of rigidity in a membrane [10]. The order parameter can vary from zero (completely fluid) to a value of one (rigid, as in a crystal). Some probes seek out the fluid parts of the membrane and can thus be used to show how much of the bilayer is in a fluid state. With such a probe, McConnell and co-workers showed that biological membranes are largely fluid [11]. A disadvantage of probes is that they report only the characteristics of the region in which they are located, which may not represent the membrane as a whole.

Recently, Chin has used EPR to study the disordering effect of ethanol in biomembranes [12]. With a sensitive instrument, she showed the effects of ordinary intoxicating concentrations of ethanol for the first time. Previous experimenters had used stronger drugs or much higher concentrations of ethanol. Ethanol added to membranes *in vitro* caused a concentration-related decrease in the order parameter, as measured with a stearic acid spin label.

Similar EPR experiments showed that inhalation anesthetics such as halothane and methoxyflurane had a disordering effect on model membranes composed of lecithin and cholesterol [13]. Two different spin-labeled phospholipids were used, with the spin label located on the fatty acid acyl chains at different distances from the head group. Figure 4-4 shows (a) that both anesthetics had the same disordering effect at the same intramembrane concentration, (b) that the deeper probe monitored a more fluid region (as expected), and (c) that the slope of the response to increasing drug concentration was the same with both probes, suggesting that the disordering effect of the drugs extended throughout the bilayer.

Another probe technique for studying the general fluidity of membranes is fluorescence polarization. Membrane samples containing a lipid-soluble

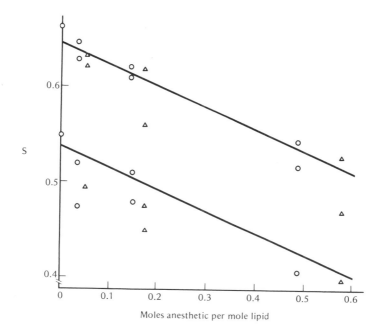

Fig. 4-4. *Disorder caused by anesthetics in model membranes.* Halothane (O) or methoxyflurane ((△) was added to suspensions of vesicles composed of egg phosphatidylcholine and cholesterol. The abscissa represents the intramembrane concentrations of the drugs, calculated from their aqueous concentrations and measured partition coefficients. The ordinate represents the order parameter, S. The preparations were spin labeled with a phosphatidylcholine that contained 5-doxylmyristic acid (top curve) or 9-doxylstearic acid (bottom curve). From Trudell et al. [13].

fluorescent dye are irradiated with polarized light. If the dye molecules were stationary, they would emit perfectly polarized fluorescence, but if they move during the few nanoseconds between absorption and emission of light, the polarization of the emitted light beam will be partially lost. By measuring the fluorescence intensity both parallel and perpendicular to the excitation polarization, one can determine to what extent the dye was free to move. This is a function of the fluidity of the lipid in which it is dissolved. It is not surprising, then, to hear that ethanol decreases polarization of fluorescence in membrane lipids labeled with diphenylhexatriene, a fluorescent dye [14].

Lipid phase transitions. Another important concept in membrane biophysics is the phase transition. Lipids can exist in the bilayers as fluids or (at lower temperatures) as gels. Phase transitions occur cleanly in model membranes with simple lipid composition, but in the complex membranes of living cells, the transition may stretch out over quite a range of temperature or may not be detectable. It appears that biological membranes, at physiological temperatures, are a mixture of fluid and solid phases [15]. There is a matrix of fluid lipid in which patches of solid are floating like ice floes. Proteins seem to be confined to the fluid domains. As the temperature changes in such a system the relative proportion of the phases will shift, with likely effects on the functions of the proteins. Changes not only in temperature but also in pH and in ionic composition can cause phase transitions that affect the degree of lateral phase separation. This is an obvious place where drugs could act.

Anesthetic drugs, including ethanol, can lower the temperature at which lipid bilayers undergo transition between the gel and liquid crystalline states. Figure 4-5 shows the lowering of the phase transition temperature by octanol in pure lipid vesicles, measured by the fluorescence of chlorophyll. The fluorescence intensity is much higher when chlorophyll is located in fluid lipid than in the gel state; thus, the abrupt increase in fluorescence intensity as the temperature is raised represents melting of the lipid. Drugs that lower phase transition temperature may increase the proportion of lipid that is fluid at body temperature—another indication of membrane disordering.

In differential scanning calorimetry, one measures the heat input necessary to raise the temperature of a sample, in comparison with a reference sample. When the sample undergoes a phase transition there will be a conspicuous change in slope of the enthalpy curve. This method also shows that alcohols lower the transition temperature.

Phase transitions can literally be seen by means of the freeze fracture technique, in which samples of membranes can be instantaneously frozen so that their molecular structure is preserved in a configuration representing their fluid or gel state. It is then possible to strike the frozen sample in such a way that it fractures between the two halves of the bilayer. With the electron microscope, the areas that were fluid or gel before freezing can be recognized [17]. With sufficient magnification, particles that are assumed to be proteins are visible in the fluid regions of the membrane. Freeze fracture has not been used to study ethanol effects, but other an-

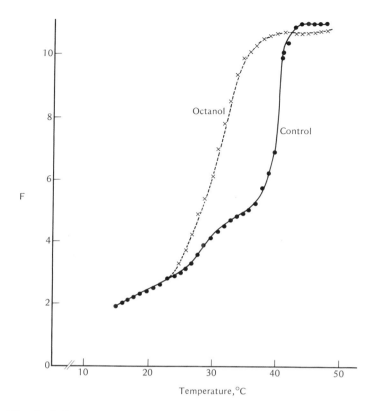

Fig. 4-5. *Alcohols can reduce the temperature of lipid phase transitions.* Octanol was added to vesicles made of dipalmitoylphosphatidylcholine with chlorophyll incorporated in the lipid. The fluorescence intensity (F) was measured as the temperature was raised. In the pure lipid an abrupt increase in fluorescence at 41° C defined the transition temperature. The "pretransition" at about 30° C, characteristic of this phospholipid but poorly understood, can also be seen here. Octanol, 2.3 mM aqueous concentration, lowered the transition temperature to about 30° C, widened the temperature range over which the transition took place, and abolished the pretransition. From Lee [16].

esthetic agents are known to increase the proportion of fluid lipid visible in freeze fracture images [18].

Arrhenius plots of enzyme action also demonstrate the phase transition. Membrane-bound enzymes function differently above and below the transition temperature, as is to be expected if their properties depend on the fluidity of their surroundings. This can often be detected by finding discontinuities in Arrhenius plots of enzyme activity. Plotting the log of the activity versus the reciprocal of the absolute temperature produces a straight line for most soluble enzymes and usually for membrane enzymes near the top of their temperature range. As the temperature is lowered, however, one may see a sharp change in the slope of the line, representing a change in the energy of activation. Such breaks may indicate the temperature of a phase transition in the region of the membrane that affects the enzyme. The effective lipid may be the bulk lipid or, more likely, the boundary lipid immediately surrounding the enzyme molecules. Many artifacts, technical barriers, and difficulties in interpretation surround the use of Arrhenius plots to observe effects of drugs on membranes, but this is a potentially useful technique.

Pressure reversal of anesthesia. Thus, a large class of chemically dissimilar drugs seems to act in accordance with lipid solubility merely by occupying space in membranes, thereby expanding and disordering them. Support for this idea came in a surprisingly straightforward way. Anesthesia can be reversed by hydrostatic pressure. Tadpoles, newts, and even mice that have lost their righting reflex after being treated with an anesthetic drug regain it under high pressure, about 200 atmos. Pressure apparently forces the membrane constituents back into order. This effect has been reproduced in an EPR study on model membranes [19]. Halothane lowered the order parameter of phosphatidylcholine-cholesterol model membranes, but high pressure partially reversed the drug effect. High pressure applied to intact animals can cause excitable states that are in some ways the opposite of anesthesia, and one can entertain, at least for the moment, the simple notion that stiffer membranes make more excitable brains. Pressure reversal of barbiturate anesthesia has been demonstrated. This means either that barbiturates act by disordering membranes or that pressure produces a nonspecific stimulant effect that can counteract any sedative drug.

Kendig has studied actions of anesthetics and pressure on the isolated rat superior cervical ganglion [20]. She found that halothane, methoxyflurane, and ethanol diminished the height of the compound action potential of the preganglionic nerve trunk. Hydrostatic pressure itself had little effect, but it did antagonize the effect of the anesthetic drugs, in agreement with the *in vivo* reversal of anesthesia. However, both pressure and anesthetics decreased synaptic transmission, either nicotinic or muscarinic. Pressure and drug effects on synaptic function were additive. This finding does not conform to current thinking about the meaning of pressure reversal of anesthesia.

Chronic effects of ethanol on membrane lipids

If ethanol acts by disordering membranes (and we do not yet know whether this hypothesis is correct), what happens after chronic exposure to ethanol? Could there be an adaptation within the lipid portion of the affected biomembranes? Hill and Bangham have suggested that membrane fluidity is correlated with the state of excitability of the brain and that adaptation to anesthetic drugs might occur by means of changes in the composition of the membrane lipids [21]. The body may respond to ethanol by appropriate changes in the chemical composition of its membranes, thus ordering them and restoring a normal viscosity in the presence of ethanol. The membranes would then be abnormally rigid in the absence of the drug, which might account for the withdrawal hyperexcitability. Some evidence can be assembled to make this an attractive idea.

Homeoviscous adaptation. Control mechanisms do indeed exist in nature for changing the fluidity of biomembranes according to environmental conditions. Experimental manipulation of fatty acid uptake in microorganisms can change the physical properties of their cell membranes and thereby change their ability to grow at different temperatures [22]. Conversely, the cells themselves will adjust their membrane fatty acid composition to suit the ambient temperature. Several investigators have shown that *Escherichia coli* incorporates more unsaturated fatty acids into its membranes when grown at low than at high temperatures [23]. This would have the effect, we assume, of maintaining the right degree of fluidity for cell functions. Sinensky has measured the relative viscosity of membrane lipids from *E. coli* cells grown at different temperatures [24]. As Table 4-1 shows, all the lipids had the same viscosity, whenever the viscosity determination was made at the growth temperature. But if the measurement was done at a temperature below the one to which the cells had adapted, the lipids were much more viscous. Sinensky has suggested a mechanism for this "homeoviscous adaptation" by showing that an acyl transferase that puts fatty acids onto the glycerol backbone changes its substrate specificity with temperature, incorporating more unsaturated fatty acid at low temperatures and more saturated fatty acid at warmer ones. Thus the evidence is strong

Table 4-1. Homeoviscous adaptation. The relative viscosity of lipids extracted from *E. coli* was measured by EPR, using a methyl stearate spin label.[1]

Growth temperature, °C	Observation temperature, °C	Correlation time, nsec
15	15	2.8
30	30	2.7
37	37	2.6
43	43	2.7
43	15	13.8

1. The relative viscosity was measured as the rotational correlation time of the spin label. Longer correlation times indicate more viscous lipids.

From Sinensky [24].

that bacteria can adapt their membrane lipids to external conditions that cause a change in fluidity.

Goldfish can also change their membrane fatty acid composition in response to the temperature to which they have been acclimated [25]. Thus we have reason to believe that organisms that have to adapt to different temperatures can do so in part by changing the composition of their membrane lipids. Homeotherms do not have to adapt their internal cells to changes in temperature, but we know they can regulate their adipose tissue lipids. Regulation of mammalian adipose tissue liquids was reported as early as 1901 [26] in a set of ingenious experiments where it was shown (a) that mammals of species that live in cold environments have more unsaturated fatty acids in their adipose tissue than warm-climate species, (b) that the adipose composition can alter with a natural change in the environmental temperature (in dolphins at birth, e.g.), and (c) that the fats can change in response to experimental manipulations of ambient temperature. This last was the famous pig-in-a-sheepskin experiment where the skin temperature was varied in three pigs by means of housing conditions or a blanket; the degree of unsaturation in subcutaneous fat varied appropriately.

Ethanol-tolerant membranes. We have recently sought direct evidence for adaptation in the membranes of mice, by examining the concentration-related disordering effect of ethanol *in vitro* in membranes from ethanol-treated mice [27]. Brain and erythrocyte membranes isolated from mice after eight days on an alcohol liquid diet were strikingly resistant to the effects of ethanol *in vitro*, in comparison with pair-fed controls (Fig. 4-6). These membranes were found to have a higher cholesterol:phospholipid ratio than the membranes of control animals [28], a condition that might help to explain the tolerance. Cholesterol buffers the effects of several membrane perturbers, including temperature and ethanol, and cholesterol can also reduce the partition coefficient of drugs in membranes.

One might expect that membranes from ethanol-treated animals would be more ordered than normal, having stiffened up to offset the disordering effect of the alcohol; this would be a form of homeoviscous adaptation. The excessively rigid membranes, being the opposite of membranes disordered by ethanol, might perhaps account for the withdrawal hyperexcitability. There does appear to be such a change under some conditions, but it is

Fig. 4-6. *Tolerance in isolated membranes.* Mice were treated with ethanol in a liquid diet for eight days; controls were pair fed an isocaloric sucrose diet. Synaptosomal plasma membranes were spin labeled with 5-doxylstearic acid and the order parameters measured by EPR in the presence of various concentrations of ethanol. Membranes from alcohol-treated mice were tolerant to the disordering effect of alcohol *in vitro*. From Chin and Goldstein [27].

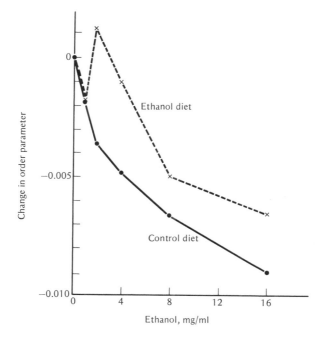

too early to say whether this is a membrane expression of physical dependence.

Summary

Ethanol fits the Meyer-Overton concept that the lipid solubility of anesthetic drugs determines their potency, which suggests that its primary action is in a hydrophobic site such as the cell membrane. Biomembranes are expanded in area and disrupted in structure by alcohols *in vitro*. Data from experiments using several physical chemical techniques agree that ethanol disorders membranes, making them more fluid and increasing the mobility of membrane components. The phase transition temperature is lowered. Alcohol may increase the proportion of the lipid that is fluid rather than in a gel state. Membrane proteins depend on the lipid for an appropriate environment, and their function may be affected by slight changes in membrane fluidity. Disruption of membrane protein function by changes in the enveloping lipid may be the mechanism for intoxication. We also speculate that when cells are exposed chronically to ethanol, they may adapt in a manner similar to temperature adaptation to keep their membranes in a suitable state of fluidity despite the presence of ethanol; this may lead to tolerance. Brain membranes from mice that have been chronically treated with ethanol contain more cholesterol than controls, and they are resistant to the disordering effect of ethanol *in vitro*. This may be a membrane expression of tolerance. An unusually rigid membrane might be the expression of physical dependence.

References

1. Meyer, K.H. Contributions to the theory of narcosis. Trans. Faraday Soc. 33:1062–1068, 1937.
2. Mullins, L.J. Some physical mechanisms in narcosis. Chem. Rev. 54: 289–323, 1954.
3. Seeman, P. The membrane actions of anesthetics and tranquilizers. Pharmacol. Rev. 24:583–655, 1972.
4. Seeman, P., Kwant, W.O., Sauks, T. and Argent, W. Membrane expansion of intact erythrocytes by anesthetics. Biochim. Biophys. Acta 183: 490–498, 1969.
5. Seeman, P. The membrane expansion theory of anesthesia: direct evidence using ethanol and a high-precision density meter. Experientia 30: 759–760, 1974.
6. Singer, S.J. and Nicolson, G.L. The fluid mosaic model of the structure of cell membranes. Science 175: 720–731, 1972.

7. Jost, P.C., Griffith, O.H., Capaldi, R.A. and Vanderkooi, G. Evidence for boundary lipid in membranes. Proc. Nat. Acad. Sci. 70: 480–484, 1973.

8. McConnell, H.M. Molecular motion in biological membranes. *In:* Spin Labeling. Theory and Applications, Berliner, L.J., Ed. Academic Press, New York, 1976. pp. 525–560.

9. Metcalfe, J.C., Seeman, P. and Burgen, A.S.V. The proton relaxation of benzyl alcohol in erythrocyte membranes. Mol. Pharmacol. 4: 87–95, 1968.

10. Hubbell, W.L. and McConnell, H.M. Molecular motion in spin-labeled phospholipids and membranes. J. Am. Chem. Assoc. 93: 314–326, 1971.

11. McConnell, H.M., Wright, K.L. and McFarland, B.G. The fraction of the lipid in a biological membrane that is in a fluid state: a spin label assay. Biochem. Biophys. Res. Commun. 47: 273–281, 1972.

12. Chin, J.H. and Goldstein, D.B. Effects of low concentrations of ethanol on the fluidity of spin-labeled erythrocyte and brain membranes. Mol. Pharmacol. 13: 435–441, 1977.

13. Trudell, J.R., Hubbell, W.L. and Cohen, E.N. The effect of two inhalation anesthetics on the order of spin-labeled phospholipid vesicles. Biochim. Biophys. Acta 291: 321–327, 1973.

14. Johnson, D.A., Lee, N.M., Cooke, R. and Loh, H.H. Ethanol-induced fluidization of brain lipid bilayers: required presence of cholesterol in membranes for the expression of tolerance. Mol. Pharmacol. 15: 739–746, 1979.

15. Shimshick, E.J. and McConnell, H.M. Lateral phase separation in phospholipid membranes. Biochemistry 12: 2351–2360, 1973.

16. Lee, A.G. Interactions between anesthetics and lipid mixtures. Normal alcohols. Biochemistry 15: 2448–2454, 1976.

17. Kleeman, W. and McConnell, H.M. Interactions of proteins and cholesterol with lipids in bilayer membranes. Biochim. Biophys. Acta 419:206–222, 1976.

18. Nandini-Kishore, S.G., Kitajima, Y. and Thompson, G.A., Jr. Membrane fluidizing effects of the general anesthetic methoxyflurane elicit an acclimation response in *Tetrahymena*. Biochim. Biophys. Acta 471: 157–161, 1977.

19. Trudell, J.R., Hubbell, W.L. and Cohen, E.N. Pressure reversal of inhalation anesthetic-induced disorder in spin-labeled phospholipid vesicles. Biochim. Biophys. Acta 291: 328–334, 1973.

20. Kendig, J.J., Trudell, J.R. and Cohen, E.N. Effects of pressure and anesthetics on conduction and synaptic transmission. J. Pharmacol. Exp. Ther. 195: 216–224, 1975.

21. Hill, M.W. and Bangham, A.D. General depressant drug dependency: a biophysical hypothesis. Adv. Exp. Med. Biol. 59: 1–9, 1975.

22. McElhaney, R.N. The effect of alterations in the physical state of the membrane lipids on the ability of *Acholeplasma laidlawii* B to grow at various temperatures. J. Mol. Biol. 84: 145–157, 1974.

23. Marr, A.G. and Ingraham, J.L. Effect of temperature on the composition of fatty acids in *E. coli*. J. Bacteriol. 84: 1260–1267, 1962.

24. Sinensky, M. Homeoviscous adaptation—a homeostatic process that regulates the viscosity of membrane lipids in *Escherichia coli*. Proc. Nat. Acad. Sci. 71: 522–525, 1974.

25. Roots, B.I. Phospholipids of goldfish (*Carassius auratus* L.) brain: the influence of environmental temperature. Comp. Biochem. Physiol. 25: 457–466, 1968.

26. Henriques, V. and Hansen, C. Vergleichende Untersuchungen über die chemische Zusammensetzung des thierischen Fettes. Skand. Arch. Physiol. 11: 151–165, 1901.

27. Chin, J.H. and Goldstein, D.B. Drug tolerance in biomembranes: a spin label study of the effects of ethanol. Science 196: 684–685, 1977.

28. Chin, J.H., Parsons, L.M. and Goldstein, D.B. Increased cholesterol content of erythrocyte and brain membranes in ethanol-tolerant mice. Biochim. Biophys. Acta 513: 358–363, 1978.

Review

Sun, A.Y. Biochemical and biophysical approaches in the study of ethanol-membrane interaction. *In:* Biochemistry and Pharmacology of Ethanol, Majchowicz, E. and Noble, E.P., Eds. Vol. 2. Plenum Press, New York, 1979, pp. 81–100.

5. Alcohol and High—Density Lipoproteins

The reader may welcome a good word about alcohol, albeit a short one. There is recent evidence that drinking increases plasma levels of high-density lipoproteins (HDL) and that people with high HDL levels have less coronary heart disease than those with low levels. The corollary that alcohol protects the cardiovascular system is not quite so clear. Light drinking may have beneficial results mediated by HDL, but heavy drinking probably damages the heart directly. Here we seek some encouraging evidence that moderate or "social" drinking may affect lipid metabolism in a desirable way, reducing the risk of atherosclerosis. The relationship between these findings and the ethanol-lipid interactions discussed in Chapter 4 is unknown, but it is reasonable to suppose that similar processes are at work. That is, alcohol-induced disruption of cholesterol-phospholipid associations, either in cell membranes or in lipoproteins, may affect the distribution of cholesterol.

Plasma lipoproteins

The atherosclerotic lesion consists of a plaque of cholesteryl esters deposited inside the cells of blood vessel walls. Eventually it may become large enough to occlude the vessel. What causes this dangerous accumulation is uncertain, but its source is probably the cholesterol in lipoproteins that circulate in the plasma. The lipoproteins are carriers for water-insoluble substances such as triglycerides and cholesteryl esters. They have an important role in delivery of cholesterol to cells, where it is used for membrane synthesis, and in the reverse process, scavenging unneeded cholesterol from tissues for transport back to the liver. Lipoprotein particles have a core of neutral lipids such as cholesteryl esters and triglycerides, and a shell of phospholipids, free cholesterol, and protein, probably in the form of a monolayer, not unlike a single leaflet of a cell membrane.

There are four main classes of plasma lipoproteins, characterized by their density or their electrophoretic mobility. Two nomenclature systems correspond to the two methods of isolation. The lipoproteins lightest in weight are chylomicrons, which are particles that carry dietary lipids from the gut to the adipose tissue, where they deposit their triglycerides, and then to the liver, where they give up their cholesterol. Their low density reflects the lowest ratio of protein to lipid of all the lipoproteins. In the commonly used electrophoresis system, they remain at the origin. The next heaviest particles are the very low density lipoproteins (VLDL), which have an electrophoretic mobility called pre-β. These particles are formed in the liver and have a high lipid:protein ratio. They proceed to the adipose tissue, where they are stripped of triglycerides by tissue lipases. The remnant is the low density lipoprotein (LDL), whose position in the electrophoretic system is designated as β. This is the particle that contains most of the plasma cholesterol and its abundance correlates closely with risk of coronary heart disease. LDL binds to specific receptors in cell membranes and is internalized and carried to the lysosomes, where it breaks down [1]. The cholesterol thus released regulates the synthesis of cholesterol in nonhepatic tissues. The heaviest of the plasma lipoproteins are the high density (HDL) particles, or α-lipoproteins. These have a lipid composition similar to LDL but contain more protein. An apoprotein of the α-lipoprotein particle activates the plasma lecithin cholesterol acyltransferase (LCAT), an enzyme that transfers acyl chains from phosphatidylcholine to cholesterol, forming lysolecithin and cholesteryl esters. The interconversions and metabolic pathways of the various lipoproteins are still unclear, but it is now thought that LDL and HDL may have complementary roles. LDL is a known risk factor for coronary heart disease and HDL may be protective. The complex interactions of the plasma lipoproteins are summarized in the review by Brown et al. that is cited at the end of this chapter.

HDL and coronary heart disease

Recent epidemiological studies show an inverse relation between HDL levels and coronary heart disease. One of several large prospective studies of risk factors in coronary heart disease is the Cooperative Lipoprotein Phenotyping Study, which includes a large sample of the general population in Framingham, Massachusetts, and a combined sample of males of Japanese ancestry living in Honolulu or San Francisco [2]. The studies have focused on serum cholesterol, which is known to be correlated with coronary heart disease, and on the LDLs, which carry about 60% of the serum cholesterol. It is reasonable to postulate that high serum cholesterol, leaking into cells whenever they are damaged (e.g., by high blood pressure), might cause the atherosclerotic lesions. Epidemiological evidence abundantly confirms the correlation of serum cholesterol with risk of coronary heart disease. However, the total plasma cholesterol does not tell the whole story. Subpopulations that have lower-than-average rates of coronary heart disease are found to have higher HDL levels. Women have higher HDL than men. Blacks have more than whites. Runners have high levels of HDL.

Multivariate statistical analyses, taking into account such obvious risk factors as smoking, diet, and lack of exercise, showed that there was indeed an inverse relationship between HDL levels and coronary heart disease. This was true both for infarction and for angina pectoris in all five of the populations in the Cooperative Lipoprotein Phenotyping Study. It was not a linear relationship, however. The increased risk was associated only with very low levels of HDL cholesterol, and there seemed to be no additional advantage in having an unusually high level (Table 5-1). It could also be shown that there was a regular increase in prevalence of coronary heart disease with increasing LDL at each level of HDL cholesterol. Clearly, total serum cholesterol measurements do not suffice for estimating risk of heart disease.

Table 5-1. Relation between HDL cholesterol levels and prevalence of coronary heart disease [1]

HDL cholesterol, mg/100 ml	Number of subjects		Prevalence rate/1000
	CHD	Total	
<25	9	50	180
25–34	78	631	124
35–44	133	1406	95
45–54	91	1168	78
55–64	45	578	78
65–74	17	215	79
>74	10	117	86
All levels	383	4165	92

1. The data are for men aged 50 to 69, reported by Castelli et al. [2].

What could be the mechanism by which HDL reduces the risk of heart disease? Glomset [3] suggested some time ago that this lipoprotein may transport cholesterol from tissues back to the liver for disposal. When cell membranes turn over, lipids must be scavenged. This can be accomplished by the HDL, through their ability to activate LCAT, thus removing free cholesterol from cell membranes and forming neutral esters in the HDL core. More recently, another protective role has been postulated for the HDL. Carew et al. [4] reported that HDL competes with LDL for binding at the LDL receptor in cultured smooth muscle cells. HDL reduced the binding, internalization, and degradation of radioactively labeled LDL in this system and thus blocked cholesterol uptake from the serum into the smooth muscle cells. Incubation of the cells with HDL alone did not affect their cholesterol content. Hence, there may be at least two mechanisms by which HDL protects against excessive tissue and plasma levels of cholesterol. The HDL may carry the excess cholesterol to the liver, thus clearing the tissues, or the HDL may impede the internalization of LDL in smooth muscle cells, thereby preventing cholesterol buildup in the vessel walls.

Alcohol and HDL

Swedish workers have found high levels of HDL in alcoholics admitted to the hospital [5,6]. Figure 5-1 shows the distribution of α-lipoprotein (HDL) levels in alcoholics on admission to the hospital. Many of the patients were intoxicated at the time of sampling. Unfortunately, many were in poor physical condition and their diets before admission were probably inadequate. Controls were healthy men. The α-lipoprotein concentration and the HDL cholesterol (both free and esterified) were much higher in the alcoholics, on the average, and were also more variable than in controls. The excess HDL disappeared in about a week of hospitalization.

Similar effects were seen in an experimental study where ethanol was administered to normal volunteers [7]. Nine healthy medical students were given ethanol for five weeks, in doses roughly equivalent to five drinks a day, spaced well apart. The total daily dose of 1.1 g/kg was not enough to elevate the blood ethanol levels above 0.2 mg/ml or to cause any signs of intoxication. Over the five-week drinking period there was a steady rise in levels of α-lipoproteins in plasma, to a peak about 30% above baseline, with partial return to normal in the four-week postdrinking period (Fig. 5-2). No significant change was seen in the β-lipoprotein fraction or in the total plasma cholesterol. Biopsies revealed a doubling of liver triglyceride content without visible accumulation of fat.

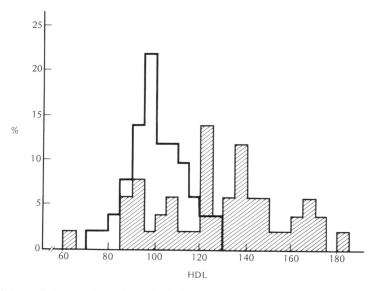

Fig. 5-1. *High levels of HDL in alcoholics on admission.* The open bars (heavily outlined) show the distribution of α-lipoprotein concentrations in serum of 50 healthy men, as percent of the sample mean. The hatched bars show the α-lipoprotein distribution in 50 consecutive male alcoholics admitted to a general hospital in Sweden, relative to the control mean. From Johansson and Medhus [5].

Recent results of the Lipid Research Clinics Program Prevalence Study [8], based at the National Institutes of Health, confirm the association between HDL levels and consumption of alcoholic beverages. In a study of 4855 white subjects [9], a dose-related increase in HDL concentrations with increasing alcohol intake was found for all age groups and both sexes. Among subjects who reported not taking any drinks during the previous week, a difference was found between teetotalers and those who sometimes drank; the latter had higher HDL levels. This suggests a persistent effect of occasional drinking. The Cooperative Lipoprotein Phenotyping Study had earlier reported similar data in their study populations, which are ethnically more diverse [10].

Alcohol and coronary heart disease

Now we come to the main question. Does moderate drinking actually improve one's chances of avoiding a heart attack? Here the evidence is not so clear, but there are suggestive data. In one study [11], the subjects were

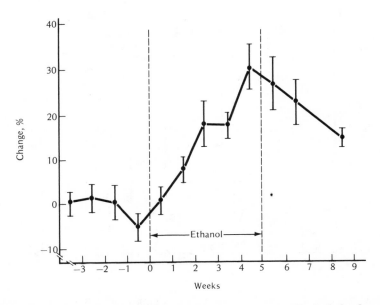

Fig. 5-2. *Increase in serum α-lipoprotein during experimental alcohol administration.* Ethanol was consumed by healthy volunteers over a period of five weeks, in the form of five nonintoxicating doses per day. Plasma α-lipoprotein concentrations are shown as the percent change from the mean baseline level, established over a four-week predrinking period. Points are means and bars are SEM for eight or nine subjects. From Belfrage et al. [7].

participants in the Kaiser Foundation Health Plan in San Francisco and Oakland. Over 460 patients were found who had had a physical exam and detailed history on entry into the plan and who had subsequently undergone a first heart attack. Two controls were found for each of these patients. One was a "normal control," matched simply for age, sex, skin color, and examination date. The other, for which the thousands of patient records in the Kaiser system were needed, was a "risk control," matched for several factors thought to contribute to risk of heart disease, such as smoking, blood pressure, and serum cholesterol. The risk controls, therefore, were those who had not had an infarction even though their smoking and cholesterol intake, etc., were as high as among those who had. The data revealed significantly more teetotalers among the infarction patients than among the risk controls. When the other risk factors were taken into account, an inverse correlation between drinking and vulnerability to myocardial infarction was found, even for the heavy drinkers.

Another study [12] yielded the same message, but with a more restricted population. In the Honolulu Heart Study, a sample of over 7000 men of Japanese descent included 294 who had a new case of coronary heart disease during a six-year period. In this group, moderate alcohol consumption clearly was associated with decreased risk of heart disease. The incidence of coronary heart disease went steadily down with increasing alcohol consumption, up to about two drinks a day (Fig. 5-3). The correlation with drinking was highly significant for total coronary heart disease and for myocardial infarction, but was not significant for deaths, coronary insufficiency, or angina pectoris. The beneficial effect of alcohol was clear even when no correction was made for smoking and was stronger when such a correction had been made. Another recent study [13] confirmed that the effect of small amounts of alcohol is independent of the type of beverage used. Heavy drinking did not affect coronary heart disease rates in either direction.

Fig. 5-3. *Decreasing incidence of coronary heart disease with increasing alcohol consumption.* Alcohol intake was estimated from interviews. The age-adjusted incidence of coronary heart disease is shown. Hatched bars represent fatal coronary heart disease and myocardial infarction; open portions of the bars indicate angina pectoris and acute coronary insufficiency. The correlations were highly significant for alcohol intake versus myocardial infarction and for alcohol intake and total coronary heart disease. From Yano et al. [12].

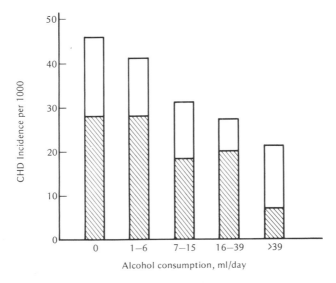

Caveat. Of course, these findings should not be construed to mean that heavy drinking is beneficial. A tremendous weight of evidence proves that heavy drinkers have shorter life spans than light drinkers. The death rates from liver disease, accidents, suicides, and many other causes add up to a poor medical prognosis for anyone who drinks heavily [14]. The data on HDL and coronary heart disease indicate that there might possibly be specific beneficial effects of drinking that are dose related and can be observed at doses too low to damage other systems. But at higher doses, toxicity to the liver and brain will come into play and the net result will definitely be harmful.

Even the cardiovascular system suffers from excessive use of ethanol [14]. Increased blood pressure is associated with alcohol intake at levels above three drinks a day, approximately [15]. Furthermore, although the coronaries may function well, the myocardium itself is apparently damaged directly by ethanol. Alcoholic cardiomyopathy is a well-known (though somewhat ill-defined) clinical entity, assumed to be due to direct effects of ethanol on the cardiac muscle [16,17]. It is not yet entirely clear whether the damage is due to alcohol per se or is in part ascribable to malnutrition (since it resembles beriberi heart disease). Microscopically, the heart may show fibrosis, edema, and enlarged, distorted mitochondria. Lipid accumulates inside the smooth muscle cells. The picture is reminiscent of the alcohol-damaged liver, but these patients seldom have advanced cirrhosis.

Summary

Epidemiologic studies reveal a clear relationship between HDL and coronary heart disease; low levels of the α-lipoprotein are associated with high rates of coronary heart disease. It is also clear that consumption of alcoholic beverages raises serum HDL concentrations. This is a progressive dose relation but does not necessarily imply a steady decrease in coronary heart disease rates, because there is apparently no increased benefit at very high concentrations of HDL. An inverse correlation between alcohol consumption and rates of coronary heart disease has also been shown in epidemiologic studies. These multivariate analyses examine the effect of ethanol itself when other factors are held constant. In the real world, the other factors may overwhelm the alcohol effect. Heavy drinkers are at risk for many other kinds of damage. The message that light drinking may be beneficial is welcome, but it should not be extrapolated.

References

1. Goldstein, J.L. and Brown, M.S. The low-density lipoprotein pathway and its relation to atherosclerosis. Ann. Rev. Biochem. 46: 897–930, 1977.

2. Castelli, W.P., Doyle, J.T., Gordon, T., Hames, C.G., Hjortland, M.C., Hulley, S.B., Kagan, A. and Zukel, W.J. HDL cholesterol and other lipids in coronary heart disease. The Cooperative Lipoprotein Phenotyping Study. Circulation 55: 767–772, 1977.

3. Glomset, J.A. The plasma lecithin:cholesterol acyltransferase reaction. J. Lipid Res. 9: 155–167, 1968.

4. Carew, T.E., Hayes, S.B., Koschinsky, T. and Steinberg, D. A mechanism by which high-density lipoproteins may slow the atherogenic process. Lancet 1: 1315–1317, 1976.

5. Johansson, B.G. and Medhus, A. Increase in plasma α-lipoproteins in chronic alcoholics after acute abuse. Acta Med. Scand. 195: 273–277, 1974.

6. Danielsson, B., Ekman, R., Fex, G., Johansson, B.G., Kristensson, H., Nilsson-Ehle, P. and Wadstein, J. Changes in plasma high density lipoproteins in chronic male alcoholics during and after abuse. Scand. J. Clin. Lab. Invest. 38: 113–119, 1978.

7. Belfrage, P., Berg, B., Hägerstrand, I., Nilsson-Ehle, P., Tornqvist, H. and Wiebe, T. Alterations of lipid metabolism in healthy volunteers during long-term ethanol intake. Eur. J. Clin. Invest. 7: 127–131, 1977.

8. Heiss, G., Johnson, N.J., Reiland, S., Davis, C.E. and Tyroler, H.A. The epidemiology of plasma high-density lipoprotein cholesterol levels. The Lipid Research Clinics Program Prevalence Study. Summary. Circulation 62: suppl. IV, 116–136, 1980.

9. Ernst, N., Fisher, M., Smith, W., Gordon, T., Rifkind, B.M., Little, J.A., Mishkel, M.A. and Williams, O.D. The association of plasma high-density lipoprotein cholesterol with dietary intake and alcohol consumption. The Lipid Research Clinics Program Prevalence Study. Circulation 62: suppl. IV, 41–52, 1980.

10. Castelli, W.P., Doyle, J.T., Gordon, T., Hames, C.G., Hjortland, M.C., Hulley, S.B., Kagan, A. and Zukel, W.J. Alcohol and blood lipids. The Cooperative Lipoprotein Phenotyping Study. Lancet 2: 153–155, 1977.

11. Klatsky, A.L., Friedman, G.D. and Siegelaub, A.B. Alcohol consumption before myocardial infarction. Results from the Kaiser-Permanente Epidemiologic Study of Myocardial Infarction. Ann. Int. Med. 81: 294–301, 1974.

12. Yano, K., Rhoads, G.G. and Kagan, A. Coffee, alcohol and risk of coronary heart disease among Japanese men living in Hawaii. New Eng. J. Med. 297: 405–409, 1977.

13. Hennekens, C.H., Willett, W., Rosner, B., Cole, D.S. and Mayrent, S.L. Effects of beer, wine, and liquor in coronary deaths. J. Am. Med. Assoc. 242: 1973–1974, 1979.

14. Dyer, A.R., Stamler, J., Paul, O., Lepper, M., Shekelle, R.B., McKean, H. and Garside, D. Alcohol consumption and 17-year mortality in the Chicago Western Electric Company Study. Prev. Med. 9: 78–90, 1980.

15. Klatsky, A.L., Friedman, G.D., Siegelaub, A.B., and Gerard, M.J. Alcohol consumption and blood pressure. Kaiser-Permanente Multiphasic Health Examination Data. New Eng. J. Med. 296: 1194–1200, 1977.
16. Regan, T.J. Ethyl alcohol and the heart. Circulation 44: 957–963, 1971.
17. Ferrans, V.J. Alcoholic cardiomyopathy. Am. J. Med. Sci. 252:89–104, 1966.

Review

Brown, M.S., Kovanen, P.T. and Goldstein, J.L. Regulation of plasma cholesterol by lipoprotein receptors. Science 212: 628–635, 1981.

6. Acute Intoxication

In this chapter we will consider quantitative methods for measuring the acute intoxicating action of ethanol in the whole animal. With these in hand, we will proceed to Chapter 7 for a discussion of tolerance.

The familiar picture of an acutely intoxicated person is actually a complex array of many different types of impairment. Many central processes are affected, including motor, sensory, and cognitive functions. Some of these can be measured in numerical terms. One should be familiar with several units of blood alcohol concentration in order to read the literature on alcohol intoxication. While biochemists sensibly use molar units throughout, pharmacologists generally use units of mg ethanol per ml blood, and clinicians express the concentrations in mg/100 ml (mg % or mg/dl). Finally, the law and the press generally use percent (by volume). As a guide to interconversion, Table 6-1 shows rough estimates of human threshold levels and lethal alcohol concentrations in the different units.

Testing for ataxia in rodents and dogs

Ataxia is a prominent component of the intoxication syndrome that is easy to reproduce in animals and can be quantitated by several simple devices. One is the tilt plane [1], a board hinged to the table top. A rat is placed on the board and the free end is slowly raised until the animal slides off. The angle is recorded, giving a graded measure of drug effect. Another device, the rotarod [2], is a horizontal dowel that rotates about its long

Table 6-1. Units of blood ethanol concentration

Minimal intoxication	Lethal
0.01 M	0.1 M
0.5 mg/ml	5 mg/ml
50 mg %	500 mg %
0.05%	0.5%

axis. Normal mice can easily walk the rod, like a lumberman in the log pond, but they will fall off if drugged. This test is usually done as a quantal measure, by noting whether each mouse can stay on the rod for some pre-set time, such as a minute. Some workers obtain graded scores with the rotarod by progressively increasing the speed of rotation until the animal falls off. In this case the recorded data are the times for individual animals.

One of the most sensitive and reliable measures [3] of ataxia in rats is their performance on a treadmill, a motor-driven belt on which they must walk a straight line, as in the traditional roadside sobriety test. A step off the belt is punished by contact with an electrified grid. The circuitry also allows recording the total time that the rat was off the belt during a two-minute test, a graded response. This is a sensitive test over a rather narrow range of blood alcohol concentrations. A disadvantage of the moving belt test is that it measures a learned behavior. The test requires expensive pretraining, and the tolerance that develops includes elements of learning, sometimes unwanted. Figure 6-1 shows a dose-response with this method.

A more subjective but successful method was used by Newman in his well-known studies of alcohol effects and tolerance in dogs, done in the 1930s [4]. Newman devised a rating system, unpretentiously named a "scale of degrees of drunkenness." With trained dogs and trained observers, the level of intoxication could be rated along a nine-point scale, from "slight ataxia on climbing stairs" all the way to coma. The degree of drunkenness was related to blood ethanol levels in the range from 2 to 4 mg/ml.

Goldberg's tests for human subjects

Motor incoordination can also be measured in humans. For this, the experiments of Leonard Goldberg in 1943 [5] have never been surpassed. He had six different tests to measure the effects of ethanol, using simple ap-

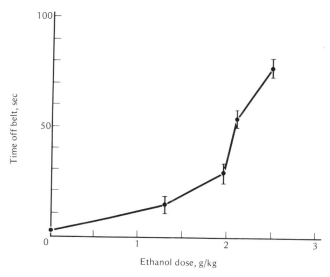

Fig. 6-1. *Moving belt test for intoxication in rats.* The animals were injected intraperitoneally with ethanol; 23 minutes later they were given three trials on the moving belt. Stepping off the belt activated a timer that recorded the "time off belt" during a 2-minute trial. The figure shows a good dose relation for moderate degrees of intoxication. Twenty rats were tested at each dose, using a Latin square design so that each rat received each dose and the order of doses was varied. Vertical bars represent SEM. From Gibbins et al. [3].

paratus. His two tests of motor incoordination were the finger-finger test and the standing steadiness test. Standing steadiness was measured by a modified Romberg test, as commonly used in neurology. The subject stood with feet together and eyes closed. The resulting body sway was recorded by a long-exposure photograph of a light attached to the subject. The area of the light squiggles, relative to each subject's baseline performance, was found to be proportional to the blood ethanol concentration. For the finger-finger test, a disk of cardboard was attached to one forefinger of the subject and a pointed thimble to the other. The subject (with arms extended) was asked to bring the forefingers together repeatedly, producing a pattern of dots on the paper. The area defined by the dots was measured with a planimeter. This area, relative to the predrug area, was found to be linearly related to the blood ethanol level (Fig. 6-2). The excellent correlation of impairment scores with blood alcohol levels in this test was shown in Chapter 2 (Fig. 2-3).

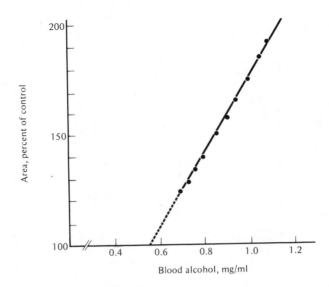

Fig. 6-2. *Concentration response in human subjects.* After ingesting ethanol, subjects were tested by the finger-finger test described in the text. The area, relative to baseline, is a measure of impairment. From Goldberg [5].

Goldberg also devised two tests of sensory perception. One was to measure the threshold intensity of a puff of air in eliciting the corneal reflex, and the other was a critical flicker fusion determination. In the latter, the subject viewed a flashing light. The frequency was increased or the intensity decreased until the subject perceived a blur instead of discrete flashes. Ethanol reduced the critical frequency or raised the critical intensity; intoxication caused the flickers to merge. In addition to his motor and sensory tests, Goldberg used two estimates of intellectual function. One of these was the marking of a specified letter whenever it occurred in a text, and the other was the "serial sevens" test, in which the subject subtracted seven successively from some starting number, usually around 100. Numerical measures of speed and accuracy were recorded for both tests. All the motor, sensory, and intellectual functions measured in Goldberg's tests showed linear relations to blood ethanol levels, when the decrement in performance was expressed as a percent of each subject's predrug score. The threshold for measurable impairment, in subjects who were habitual light drinkers, was 0.5–0.9 mg/ml in each of the tests.

Judgment and risk-taking

In administering these tests, Goldberg noted that subjects were often unable to judge the degree of their own intoxication. When their performance was obviously impaired, they often volunteered the comment that they were doing especially well. This impairment of judgment is common and it can obviously be dangerous, especially in relation to skilled behavior such as driving. Another group of investigators later designed an ingenious experiment to measure directly the recognition of hazard by intoxicated drivers [6]. Subjects for this experiment were experienced bus drivers, each of whom had received an award for safe driving. The test involved driving a bus between two stakes, which were initially set too close together. The driver was first asked whether he thought he could drive the bus through the gap, and the distance between stakes was adjusted until the answer was affirmative. Then the driver was asked actually to attempt driving through, and the size gap through which he could in fact drive the bus was recorded, again by widening the gap if necessary. Different groups of drivers were given 2 oz. or 6 oz. of whiskey, or none at all, about half an hour before the test. The results were as one might expect. Drinking increased the size of the gap that was necessary for them actually to drive through, but it decreased the size gap they would attempt. Alcohol damaged performance while increasing the driver's estimate of his ability. A limitation of this experiment was that it did not provide adequate motivation for safe driving. Although these drivers had had no accidents when carrying passengers through crowded city streets, subjects in all three groups indicated willingness to drive the bus through gaps that were actually narrower than the bus itself. Nevertheless, the effects of ethanol are clear and the effect being demonstrated certainly contributes to the damage done by drinking drivers and industrial workers.

Overall driving ability

Another driving study of an entirely different design gives us the best available dose relation with respect to traffic accidents. This was a huge study [7] done in Grand Rapids, Michigan, by a research team in cooperation with the local police. About 6000 drivers who were involved in traffic accidents were interviewed on the spot and their blood alcohol was measured. A control sample of over 7000 drivers was matched for site of the accident,

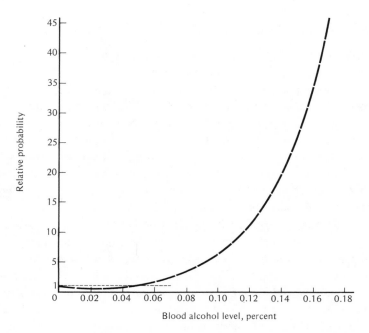

Fig. 6-3. *Relative probability of causing an accident at different blood alcohol levels.* Thousands of drivers were stopped at sites where accidents occurred, and their blood alcohol was measured. Some had actually been in accidents, and some had not. The probability of being in an accident increased sharply with blood alcohol concentration. From Zylman [8].

time of day, and day of the week. These drivers were stopped for interviews and for blood samples without having been in an accident. From these data, one could calculate the relative probability of causing an accident at different blood alcohol concentrations. Figure 6-3 shows the results. The probability goes up more and more steeply as the blood alcohol rises. No effect can be seen below 0.04% (0.4 mg/ml). Some observers note the very small apparent improvement at very low blood ethanol concentrations and wishfully believe that driving may be safer after a small drink. This is a doubtful effect at best and not to be counted on. It would be all too easy to overshoot into the steep part of the curve. The curve shows that the probability of causing an accident is doubled at a blood level of 0.6 mg/ml; it is increased six times at 1 mg/ml and over 25-fold at 1.5 mg/ml. In most states a driver is not presumed drunk unless the blood alcohol is above 1 mg/ml, a value that is already quite dangerous according to the Grand Rapids study. There are still a few states where the obviously damaging level of 1.5 mg/ml must be reached before the law considers the individual to be clearly under the influence.

The tests described above were used in classic experiments and some of them are the basis for the most reliable measures of alcohol tolerance, which is the theme of the next chapter. An enormous variety of other tests are also in use, including reaction time, pursuit rotors, and driving simulators. These can be used in human subjects to study many psychoactive drugs and CNS depressants. Factors that cause difficulties in their use and interpretation include variable motivation and the effects of practice, anxiety, and boredom. A thorough review of these methods by Hindmarch is cited at the end of this chapter.

Summary

This chapter describes some useful tests for acute intoxication in laboratory animals and in human subjects. In animals, ataxia is the easiest to measure. In people, ataxia, blunted sensory perception, and impairment of simple intellectual function can be assessed. Tests of driving skill show that small doses of ethanol impair judgment of risk, as well as actual performance, and that traffic accidents increase with blood ethanol levels at concentrations well below the legal criteria for drunk driving.

References

1. Arvola, A., Sammalisto, L. and Wallgren, H. A test for level of alcohol intoxication in the rat. Quart. J. Stud. Alc. 19: 563–572, 1958.
2. Dunham, N.W. and Miya, T.S. A note on a simple apparatus for detecting neurological deficit in rats and mice. J. Am. Pharmaceut. Ass. 46: 208–209, 1957.
3. Gibbins, R.J., Kalant, H. and LeBlanc, A.E. A technique for accurate measurement of moderate degrees of alcohol intoxication in small animals. J. Pharmacol. Exp. Ther. 159: 236–242, 1968.
4. Newman, N.W. and Lehman, A.J. Nature of acquired tolerance to alcohol. J. Pharmacol. Exp. Ther. 62: 301–306, 1938.
5. Goldberg, L. Quantitative studies on alcohol tolerance in man. Acta Physiol. Scand. 5: suppl. 16, 1943.
6. Cohen, J., Dearnaley, E.J. and Hansel, C.E.M. The risk taken in driving under the influence of alcohol. Br. Med. J. 1: 1438–1442, 1958.
7. Methodology of the Grand Rapids study. Quart. J. Stud. Alc., Suppl. 4: 267–269, 1968.

8. Zylman, R. Accidents, alcohol and single-cause explanations. Quart. J. Stud. Alc., Suppl. 4: 212–233, 1968.

Review
Hindmarch, I. Psychomotor function and psychoactive drugs. Br. J. Clin. Pharmacol. 10: 189–209, 1980.

7. Tolerance

In this chapter we discuss the magnitude, dose response, and time course of tolerance to ethanol, as manifested by behavioral effects in whole animals. Drug tolerance is defined as a diminution of a drug effect after a period of administration of that drug. This definition only applies to changes that arise as a result of exposure to a drug and does not include the many differences between individuals that cause them to respond differently to a drug on their first encounter with it. Species, sex, age, genetic background, and disease states may affect an individual's sensitivity to a pharmacological agent. By definition, tolerance only arises after exposure to the drug or to a closely related drug (cross-tolerance). Tolerance can be of two main types, dispositional or functional.

Dispositional tolerance

Dispositional tolerance is simple enough. The drug becomes less effective after chronic use because there is less of it at its site of action. Theoretically, dispositional tolerance could arise because of slower absorption from the stomach or entry into the brain, but no such effects are known. The main, if not the only, component of dispositional tolerance to ethanol is metabolic tolerance. The rate of metabolic inactivation of the drug is increased after chronic administration; thus, a particular dose produces lower and shorter-lasting blood levels.

The magnitude of metabolic tolerance to ethanol is never very large (30–

50% after experimental ethanol administration in animals), and in some experiments it is not seen at all. The mechanism could involve any of the components of ethanol metabolism that were described in Chapter 1. Liver alcohol dehydrogenase activity sometimes increases after chronic alcohol intake, but not always. Even when it increases, it might not produce an increased rate of alcohol elimination, since the rate of NAD regeneration might not be enough to keep pace with the extra enzyme. The enzymes responsible for NAD regeneration may themselves be the mechanism of metabolic tolerance. The most active dehydrogenases are those that use NADH to produce lactic, malic, and β-hydroxybutyric acids from their oxidized forms. MEOS, like most microsomal enzyme systems, is inducible, and augmentation of ethanol oxidation in liver microsomal fractions has been convincingly documented in animals after chronic treatment with ethanol [1]. There is also an increase in hepatic smooth endoplasmic reticulum, visible by electron micrography, and in the amount of cytochrome P-450. Because we do not know what proportion of *in vivo* ethanol elimination is attributable to MEOS, it is not clear how to interpret the induction.

Metabolic tolerance has been difficult to demonstrate, partly because of its small magnitude and partly because it seems to vary according to other influences on the liver. Changes in nutritional state during long-term experimental ethanol administration may explain some of the discrepancies. In alcoholics, the liver may eventually become so badly damaged that there is a decrease, rather than an increase, in its ability to handle ethanol. Thus, we are fairly sure that a moderate amount of metabolic tolerance to ethanol can develop, but we are quite uncertain as to its actual occurrence in a given patient or experiment.

Functional tolerance

Functional or tissue tolerance denotes an actual change in the sensitivity of the brain to a given concentration of the drug. Included in this definition are some types of behaviorally augmented or conditioned tolerance. The various forms of tolerance have different mechanisms and presumably different sites, time courses, and dose relations, even though all have the same ultimate effect in that the subject appears less intoxicated after chronic alcohol administration. This overall change is of practical importance to drinkers, but it tells us little pharmacologically, until we can dissect out its various components.

Measurements of the magnitude of tolerance. Functional tolerance is best demonstrated by a shift to the right in a concentration-response curve. Note that we measure the effect of a given blood concentration, rather than dose, because we must rule out any form of dispositional tolerance that results in lowered blood levels at a particular dose. Characteristics of functional tolerance that are of interest are its magnitude, its time course and reversibility, and the amount of exposure to alcohol that is necessary to evoke it.

An experiment done in 1938 by Newman and Lehman [2] illustrates the shift in response to ethanol. Dogs were given ethanol in their drinking water chronically, in such a way as to assure that high blood ethanol levels would be attained. Fluid was available only for an hour in the morning and again in the evening, so the dogs drank enough to become noticeably intoxicated twice a day. Tolerance was shown as a shift in the concentration-response curve to the right (Fig. 7-1). Impairment was established by the "scale of degrees of drunkenness" described in Chapter 6. To estimate the magnitude of tolerance numerically, one could compare the curves in either the vertical or horizontal direction. Doing it vertically, we could take a given drug concentration and see what degree of drunkenness is reached in the naive and the tolerant subject. This only works when we can compare different levels of impairment numerically, and it is not appropriate in this experiment because a dog that cannot stand (Newman's level 6) cannot be described as being exactly twice as drunk as a dog that cannot climb stairs (level 3). Generally, it is better to compare concentration-response curves in the horizontal direction. Here we take a specific endpoint—it can be any drug effect that is reliably measurable—and determine the blood level at which that endpoint occurs in the naive and tolerant state. Then we can define the tolerance as the increase in drug concentration necessary to achieve the specified endpoint. In Figure 7-1, note that a score of 3 was attained by normal dogs at about 2.9 mg/ml and by habituated dogs at about 3.8 mg/ml—a 30% increase.

Leonard Goldberg used his reliable behavioral tests (Chapter 6) to compare three groups of subjects: heavy drinkers, moderate drinkers, and abstainers [3]. In all six tests, subjects with a history of heavy drinking were less affected than moderate drinkers, and abstainers were the most sensitive (Fig. 7-2). Abstainers and heavy drinkers differed in their response to ethanol by a factor of 1.5–2 in blood alcohol concentration. Goldberg quite reasonably interpreted this difference as tolerance, i.e., the result of different drinking histories. In this experiment, however, no before-and-after

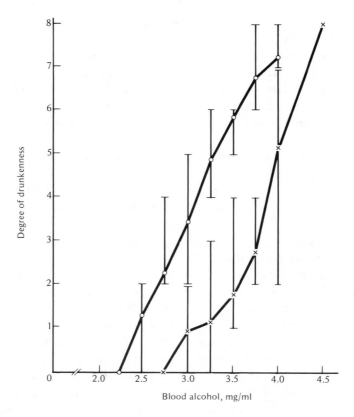

Fig. 7-1. *Alcohol concentration-response curves in normal and tolerant dogs.* Five trained dogs were tested for alcohol effects, using Newman's behavioral scale, before (o) and after (x) a three-month period of alcohol administration, during which they were intoxicated twice daily. Points are means and vertical bars show the nonoverlapping range of scores for the individual dogs. The shift to the right in the concentration-response curve represents functional tolerance. From Newman and Lehman [2].

data could be gathered in the same subjects. Thus it is not certain that the observed differences in sensitivity to ethanol were acquired as a result of previous drinking. It is surely conceivable that the subjects differed in innate sensitivity to ethanol and that the most sensitive became the lightest drinkers.

Goldberg's are the best available data on the capacity for alcohol in habitual drinkers; it is at most twice that of nondrinkers. Other drugs, especially opiate narcotics, can elicit much more tolerance than this.

Time course of onset of tolerance

Goldberg was dealing with presumed tolerance arising over the course of years of drinking, and Newman exposed his dogs to ethanol for a few months. Both were in accord with the thinking of their time. Alcoholism is seen in middle-aged people after years of heavy drinking, and for this reason it was assumed that tolerance took a long time to develop. But more recent evidence has overturned that concept. Several investigators have shown that substantial tolerance can develop within a few weeks, days, or even hours. Eugene LeBlanc and co-workers have done the most extensive studies, using their moving belt system described in Chapter 6.

Chronic tolerance. In studies of tolerance development in rats over a few weeks' time [4], LeBlanc administered ethanol by intubation at increasing daily doses from 3 to 9 g/kg. Every three days, the rats were challenged with ethanol at three different doses and tested on the moving belt. (The remainder of the daily maintenance dose was given after the test.) Controls

Fig. 7-2. *Alcohol concentration-response curves in humans, according to drinking history.* Impairment was measured by the finger-finger test (Chapter 6) in groups of abstainers (left-hand curve), light drinkers (middle curve), and heavy drinkers (right-hand curve) at known blood alcohol concentrations. Goldberg interpreted these nonoverlapping curves as evidence of tolerance. From Goldberg [3]

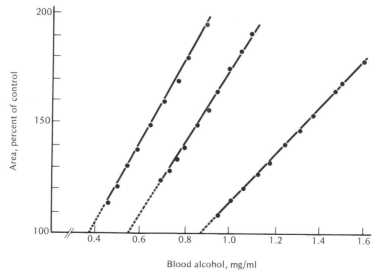

received no ethanol except the small test dose every three days. The dose-response curves of the alcohol-treated and control rats separated within a few days as the ethanol group became less and less impaired by the challenge dose. They remained apart until the end of the 24-day treatment period. Thereafter, they drifted back together. The groups clearly still differed at six days after stopping the ethanol, but not at ten days. Both acquisition and loss of tolerance can occur in a week or two. Although blood concentrations were not measured in this experiment, it is clear that the tolerance being measured was not metabolic because the tests were carried out within 30 minutes after intraperitoneal alcohol injection, at a time when elimination rates could have little effect on the blood concentration. The rats were tested repeatedly after each challenge dose of ethanol, which allowed them to practice belt-running while drunk. We will come back to this point later on, as Leblanc did.

The Mellanby effect and the distribution artifact. Not only can tolerance be demonstrated after only a few weeks of alcohol administration, but it is now clear that tolerance also can develop within a few hours, i.e., during a single drinking episode. It has been known for decades that the effects of ethanol at a given blood level are more pronounced when the blood alcohol level is rising than later, when it is falling. As Mellanby reported in 1919 [5], "A dog begins to show signs of intoxication when the alcohol of the blood reaches about 354 cm. [*sic*] per 100 grammes of blood. At this stage it will probably hit its hind toes against the floor on walking. Its movements will be slower. . . . When the alcohol has declined again to 354 cm. the dog will probably appear almost normal, and will certainly be less intoxicated than at the corresponding point on the ascent of the curve." Part of this effect has since been shown to be an artifact caused by a mistaken assumption that the concentration of ethanol in the brain is the same as in the blood. To the contrary, during the first 15 to 20 minutes after administration, the peripheral venous blood, which is usually the source of the sample taken for analysis, is losing its alcohol to the tissues and contains less ethanol than the arterial blood. But the brain, with its large blood supply, equilibrates immediately with arterial blood. Consequently, during absorption the brain alcohol concentration is quite a bit higher than is indicated by sampling the venous blood. This was discussed in Chapter 1 and illustrated in Figure 1-3. Inspection of the figure reveals that a mouse in this experiment might be in a specific behavioral state at two minutes after injection with a venous blood ethanol level (rising) of 0.2 mg/ml and might reach the same state again two hours later with the venous blood alcohol (falling) at 1.3 mg/ml. This would look like the rapid development of tolerance but could be explained better by the fact that the brain alcohol concentration was the same (1 mg/ml) at both observation times. This artifact is part of the reason for the Mellanby effect.

Acute tolerance. Since this distribution artifact was noted, however, investigators have taken it into account and they still find evidence for acute tolerance. For example, Maynert and Klingman [6] noted that dogs sober up at higher blood alcohol levels after high doses than after low doses. The whole comparison was done long enough after drug administration that the brain and blood concentrations were the same. The blood samples were taken from the jugular vein to further guard against artifact. The higher dose naturally took longer to wear off, and the measured endpoint (a certain level of ataxia) thus occurred an hour or two later in the dogs given the higher dose. Apparently this short time was sufficient for considerable tolerance to arise; blood levels at threshold were approximately twice as high after large doses than after small ones. Maynert and Klingman considered that the peak brain levels (or maximum intensity of effect) determined the magnitude of tolerance, but they could not clearly establish the role of duration of exposure to ethanol.

More recently, LeBlanc and colleagues have done a fine experiment on acute tolerance, with startling results [7]. Again using the moving belt test, they used each trained rat only once, in a two-minute test beginning exactly at 9, 29, or 59 minutes after an intraperitoneal injection of ethanol. Doses and times were arranged so as to have all the rats within a measurable range of impairment at the specified time. Distribution was not a factor since brain ethanol (not blood) was measured, and practice was not involved since the rats were confined to small cages after injection. Immediately after each test, the rat was sacrificed and the brain concentration of ethanol was measured. The results clearly show three separate and parallel dose-response curves, at the three times (Fig. 7-3). The animals were much less impaired at a given brain level of alcohol at 60 minutes than at 10 minutes after injection. What is surprising about this experiment is the magnitude of the tolerance. Comparing brain levels at a moderate level of impairment (time off belt), we have about a two-fold tolerance, arising in an hour. This is a greater magnitude of tolerance than was observed in LeBlanc's rats after four weeks of ethanol administration, and is about the same as the amount of tolerance that developed in Newman's dogs after a

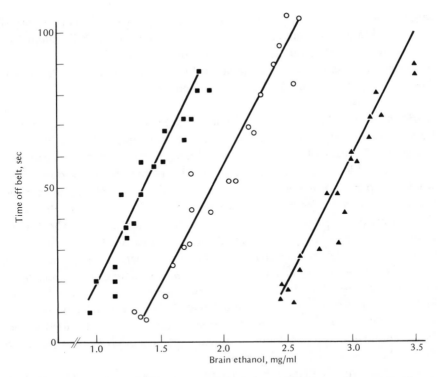

Fig. 7-3. *Acute tolerance in rats.* Points are brain alcohol concentrations and corresponding degree of impairment in the moving belt task in individual rats tested at 10, 30, or 60 minutes after an intraperitoneal injection of ethanol. Symbols are squares, 10 min; circles, 30 min; triangles, 60 min. The shift of the concentration-response curve to the right between 10 and 60 minutes represents substantial tolerance. From LeBlanc et al. [7].

few months and in Goldberg's subjects over years of drinking (cf. Fig. 7-2). Are we seeing here the rapid beginnings of a continuous process that has already reached its maximum after an hour, or are there different mechanisms for acute and chronic tolerance, with the first subsiding as the other arises?

Homeostat hypothesis

As a working hypothesis on the mechanism of functional tolerance, we may consider it as an adaptive process. The actions of alcohol are offset by the development of tolerance, much as the effects of environmental changes are offset by physiological adaptations. Perhaps the animal adapts to the presence of a drug in its internal environment by the same sorts of mechanisms it uses to adapt to external changes. What would be the time course of such a process? Adaptation to cold or hypoxia occurs by multiple responses with different mechanisms and time courses. The first response to a cold environment occurs within seconds and is neuronal in nature, consisting of shivering and piloerection. This quickly arising effect subsides rapidly if the animal returns to a warm place. But that is not all that happens. Other responses, presumably beginning as soon as the cold is encountered, take longer to peak. Enzymes for heat production are induced (nonshivering thermogenesis). The halftime for such a process is simply the half-life of the protein molecule, usually a matter of several hours or a few days. Similarly, on return to the warm, the rate of enzyme synthesis will return to normal and the excess enzyme will disappear with a halftime of hours or days. Finally, slow adaptive processes involve actual growth of tissues. In the cold, for example, there is hypertrophy of the thyroid and of the adrenals, as well as changes in the skin. These can take weeks to appear and weeks to regress.

Are there analogous telic responses to ethanol or other chronically administered drugs? No such mechanisms have been discovered as yet, but the rapid onset of tolerance and its persistence throughout the period of ethanol administration suggest that there might be different types of adaptation needed for different situations. Possibly the acute tolerance, like shivering, arises quickly and later dies away when there has been sufficient development of a longer-term biochemical adaptation.

Behavioral augmentation of tolerance

Tolerance might therefore be quite a complex phenomenon, with several different mechanisms operating simultaneously. Besides the straightforward physiological mechanisms, there is a form of tolerance in which practice and learning are involved. A clear demonstration of this was an experiment [8] in which rats were trained to run from an entry alley into a circular maze, go twice around, and reenter the alley for a food reward. After training to stable behavior, the rats were run daily in 10-minute test sessions. Photoelectric cells recorded the number of total runs—a measure of general activity—as well as the number of correctly performed runs. Every third day was a test day on which two groups of rats were treated differently. The "behavioral group" received alcohol (1.2 g/kg) 10 minutes before running the maze. The "physiological group" received the same dose after, rather than before, the test session. Thus the behavioral group had the opportunity to practice running the maze while drunk, but both groups received the same amount of drug and the same number of trials in the maze. If tolerance could arise willy-nilly according to the exposure of the brain to alcohol, then it should be the same in both groups. In the fourth and final test session, all rats received ethanol before testing, for comparison of performance between groups. Control data are combined results of interspersed sessions where saline was given instead of ethanol; rats of the two groups behaved the same in control sessions. The results (Table 7-1) show severe impairment on the first day in the behavioral group (the only group in which it could be tested), with a progressive rise in scores through the next three test days until nearly normal performance was achieved (26 correct runs). By contrast, the physiological group, which was tested only on the final occasion, still performed very badly (only three correct runs). This group had not acquired tolerance, despite having had the same doses of alcohol as the behavioral group. The lack of tolerance in the physiological group is not unexpected because the alcohol administration regime was mild, consisting of four small doses of ethanol at three-day intervals. But even this temperate regimen produced tolerance in the behavioral group. Further, the learned tolerance seemed to be specific for the maze task. The right-hand column in Table 7-1 shows that tolerance to the general depressant effect of ethanol (total trials) had developed to the same extent in both groups.

LeBlanc's lab was quick to follow up [9]. Confirming Chen's experiment,

Table 7-1. Chen's demonstration of a practice effect in the development of tolerance

	Test no.	Median no. of trials	
		Correct	Total
Control		30	35
Behavioral	1	0.5	8
	2	9	29
	3	24	33
	4	26	34
Physiological	4	3	28

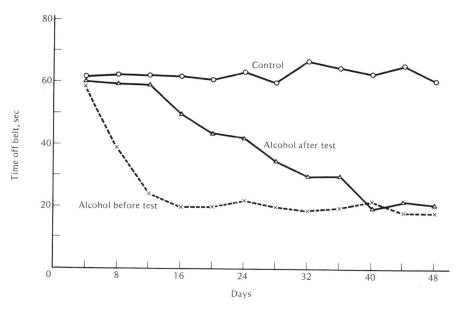

Fig. 7-4. *Behavioral augmentation of tolerance.* The curves show impairment (time off belt) of rats given a challenge dose of ethanol just before testing, every fourth day. The three groups differed in the treatment they received, along with daily trials, on the three intervening days of each four-day period. The controls received no alcohol on these days. The rats shown in the middle curve received alcohol after the trials, and those in the lower curve were given the alcohol before being tested. Only this latter group practiced daily while intoxicated, but the "before" and "after" groups received the same amount of alcohol. Tolerance developed in these two groups, but it developed faster in the rats that practiced. From LeBlanc et al. [10].

they showed that a behavioral group developed tolerance quickly in the maze task, at a time when the physiological group was almost unaffected. But in a further two weeks of continued testing, the physiological group caught up. Both groups were then at the same level of tolerance, which could not be increased by raising the alcohol maintenance dose. Thus it appeared that the behavioral and physiological groups differed only in the rate at which tolerance developed. A similar experiment using the moving belt test is shown in Figure 7-4. LeBlanc and Kalant call this phenomenon "behavioral augmentation of tolerance," and they consider that the stimulus for the development of tolerance is the impairment of some ongoing brain function. The tolerance was graded, increasing with the amount of practice under ethanol, and was reversible [10]. When the alcohol was discontinued, the rats lost their tolerance at a rate that did not depend on the frequency of (now sober) testing.

Another experiment of an entirely different nature also indicates that only functional systems develop tolerance [11]. In an identified synapse in the abdominal ganglion of *Aplysia,* addition of ethanol (0.8 M!) accelerated the decay of posttetanic potentiation. Tolerance developed over several hours of intermittent testing, but only if the cell was stimulated during the application of ethanol. Interpreting these experiments speculatively in biochemical terms, one might imagine that functioning synapses may turn over their cell membrane components more rapidly than quiescent synapses, allowing an opportunity for an adaptive change in chemical composition.

The notion that impaired function is the trigger for the development of tolerance implies that tolerance will not arise in systems that are not affected by the drug dose used during acquisition of tolerance. This idea is supported by elegant experiments of Okamoto and Boisse [12], using pentobarbital in cats. Tolerance was tested after chronic administration of low or high doses of pentobarbital and was shown as a shift in the blood level at which a particular behavioral state was produced. Tolerance developed only to those states that were produced during the chronic administration at the dose used.

State-dependent learning. We now consider experiments under the rubric of state-dependent learning, or dissociated learning. Learning in one state is said to be "dissociated" from learning in another state. Animals can best recall something when they are in the same state, e.g., drug or no-drug state, in which they learned it. This can be demonstrated in a 2×2 design. Four groups of subjects are trained, either drugged (D) or not (N), on day 1 and then tested for recall, again drugged or not, on day 2. The four groups are designated NN, ND, DN, and DD. A typical experiment with this design, using alcohol in human subjects, was carried out by Goodwin et al. [13] (Table 7-2). They used 48 male medical students for a series of five tests carried out after drinking either a soft drink or a stiff dose of vodka. The results were clearest with the two tasks shown in the table. In the rote learning task, subjects memorized a few short sentences, some of which were nonsense. They were tested on the day of learning and again the next day in the same or a different drug state. On day 1 the sober groups (NN and ND) had fewer errors than those who were drunk (DN and DD). On day 2, those who had learned when sober recalled the sentences with only a few additional errors, even if tested after drinking. How-

Table 7-2. Goodwin's experiment on state-dependent learning[1]

	Mean errors		Word association
	Rote learning		
	Day 1	Day 2	Day 2
NN	12	14	1.2
ND	12	15	2.2
DN	21	25	4.6
DD	17	16	2.5

1. Standard errors were about 2 for the rote learning test and 0.4 for the word association test. Data from Goodwin et al. [13].

ever, the remarkable result is in the second-day data for the subjects who learned after drinking. The group (DN) that was tested sober had forgotten some of the material the next day (25 errors), but the group that learned and recalled while drunk (DD) did significantly better (16 errors). That is, they actually performed better drunk than sober if they had learned while drunk. A corresponding result happened in the word association test. In this test, the subjects were presented with a set of words on day 1, and they responded to each word with another of their own choice. On day 2, they were given the same cues and asked to recall the associated words. Thus there are test results only for day 2 in this test. Again the better performance (fewer errors) was recorded by the groups that learned and recalled in the same state. The variables, including task specificity, that affect state-dependent learning are poorly understood, and the phenomenon is not yet well characterized. It is probably related to the alcoholics' claim that they must get drunk again to find the bottle they hid during a previous binge.

Drug discrimination. Drug discrimination [14] is a form of state-dependent learning that allows the animal to discriminate between two states and to tell us by its behavior that it can do so. A rat is placed in a T-maze with an electrified grid floor. It can be trained to escape by turning left or right according to its drug state, *i.e.*, turning one way after saline injections and the other way after a drug. The speed with which the animal learns to tell whether it has received the drug (the speed with which the drug acquires discriminative control) is taken as a measure of the degree of dissociation between the two states. For example, it was shown that rats learn within a few trials to discriminate any of several sedative drugs from the non-drugged state. Ethanol was one such drug. Barbiturates, meprobamate, and urethane could also control the behavior with about the same time course. Furthermore, one can in effect ask the rat whether one drug feels like another. The animal trained to turn right with ethanol and left with no drug has no trouble turning right when given a dose of another depressant drug such as pentobarbital. However, the ethanol and barbiturate states are not identical. Rats can learn to tell them apart if trained on ethanol versus barbiturates rather than on drug versus no drug. Even though alcohol and barbiturates have similar pharmacological effects, the rat's ability to discriminate between them indicates that their effects are not identical. The phenomenon makes good pharmacological sense. Animals can discriminate between drugs of different classes. They can distinguish atropine from pentobarbital, for example, but not atropine from scopolamine. Drug antagonists have predictable effects; e.g., bemegride, a mild CNS stimulant, will prevent the rat from responding to pentobarbital. Overton has observed that among the dozens of drugs tested, those that rats can discriminate are the drugs that people tend to abuse. He believes that the drugs may somehow allow the brain access to a repertoire of functions that have been learned in a drug state and cannot otherwise be experienced. In some way that is still mysterious, this may be related to addictive behavior.

Tolerance as conditioning. Let us return to tolerance, to add one more layer of behavioral complexity. Siegel has studied the situation specificity of tolerance as an example of Pavlovian conditioning [15]. He hypothesizes that the drug effect evokes physiological mechanisms that tend to counteract it (as hypothermia might elicit shivering) and that these adaptive processes become conditioned to environmental cues. The repeated association of the drug effect with a certain environment and injection ritual leads to a state where the environment itself evokes the adaptive response, so that the drug effect is attenuated. Giving a placebo at this point should bring on a state opposite to the original drug effect.

An experiment with morphine illustrates conditioned tolerance. Two groups of rats were made tolerant to morphine by repeated injections, using distinctly different environments (different rooms and noise levels), as well as different tests for analgesia. Then the rats were switched to the opposite environment and given both tests. Tolerance was seen only in the environment that matched the tolerance acquisition conditions, suggesting that the testing environment must have contributed appreciably to the tolerance. This goes beyond behavioral augmentation of tolerance, because in this case it is not even necessary for the rat to do anything in order to develop tolerance; practice in the analgesia tests was shown to be irrelevant. Apparently the receipt of environmental cues, rather than any active response of the rat, stimulates the development of tolerance.

Cross-tolerance

Cross-tolerance means that exposure to one drug reduces the subject's sensitivity to another drug. Because drug tolerance can arise by many mechanisms, a great number of such interactions is possible. Metabolic cross-tolerance is common since drug administration often induces the liver mixed-function oxygenase system and these enzymes have a limited specificity. Once induced, they will accelerate the metabolism of several other drugs. Chronic administration of ethanol may induce MEOS activity and thus produce metabolic cross-tolerance to other drugs that are degraded by the same enzymes, such as barbiturates. The reverse process is not so likely to be observed; even if MEOS were induced after administration of another drug, its effect on alcohol elimination *in vivo* would be small because alcohol dehydrogenase predominates in control of ethanol elimination. There is little evidence for induction of alcohol dehydrogenase by ethanol or any other drug.

If two drugs share a metabolic system, they may interact in a different way, with an effect that is the opposite of cross-tolerance. Whenever both are present together, they compete for the enzyme and block each other's breakdown. This would increase the acute effects of both drugs. Both cross-tolerance and competition apparently occur between ethanol and acetaminophen, an analgesic drug. Acetaminophen has a toxic metabolite that can damage the liver, sometimes with fatal results. While ethanol is being metabolized by MEOS, it competes with acetaminophen for enzyme sites and thus protects the liver. Conversely, after chronic alcohol intake and induction of microsomal enzymes, the liver is particularly vulnerable to acetaminophen [16].

Functional cross-tolerance is surprisingly difficult to demonstrate. Many pharmacologists assume that cross-tolerance exists between ethanol and barbiturates, but the data are inconsistent and sometimes negative. Currently it is believed that barbiturates may act at a receptor in the GABA receptor–benzodiazepine binding site–chloride ionophore complex in neuronal membranes, rather than acting nonspecifically in membranes like ethanol [17]. If this is so, one would not expect cross-tolerance to develop readily between alcohols and barbiturates. Cross-tolerance between ethanol and inhalation anesthetics does seem to exist, confirming clinical impressions. Cross-tolerance between ethanol and halothane in mice has recently been reported [18], and it was ascribed to a change in membrane lipids such that the partition coefficients for both drugs were reduced. Strictly speaking, this would not be functional tolerance but a molecular form of dispositional tolerance—the drugs are partially excluded from their site of action in the tolerant membrane. Cross-tolerance between ethanol and other aliphatic alcohols has been reported several times.

Summary

Tolerance to ethanol has many components that are being unraveled in the laboratory. Dispositional or metabolic tolerance is simply a matter of faster elimination of the drug after its chronic administration. Functional tolerance, which is a decreased sensitivity of the brain to ethanol, has several mechanisms and time courses. It is usually measured as a change in sensitivity over weeks of drug administration, but we now know that it can arise in hours. We do not know whether the mechanisms for acute and chronic tolerance are the same, but we speculate that physiological adaptive processes may be involved. Several kinds of evidence suggest that only functioning neuronal systems develop tolerance.

Metabolic cross tolerance is fairly common between alcohol and other drugs, including barbiturates, but functional cross tolerance seems to occur only among the drugs that act in cell membranes.

References

1. Lieber, C.S. and DeCarli, L.M. Hepatic microsomal ethanol-oxidizing system. In vitro characteristics and adaptive properties in vivo. J. Biol. Chem. 245: 2505–2512, 1970.
2. Newman, H.W. and Lehman, A.J. Nature of acquired tolerance to alcohol. J. Pharmacol. Exp. Ther. 62: 301–306, 1938.
3. Goldberg, L. Quantitative studies on alcohol tolerance in man. Acta Physiol. Scand. 5: suppl. 16, 1–128, 1943.
4. LeBlanc, A.E., Kalant, H., Gibbins, R.J. and Berman, N.D. Acquisition and loss of tolerance to ethanol by the rat. J. Pharmacol. Exp. Ther. 168: 244–250, 1969.
5. Mellanby, E. Alcohol: its absorption into and disappearance from the blood under different conditions. Med. Res. Comm., Special Report Series, No. 31, 1919.

6. Maynert, E.W. and Klingman. G.I. Acute tolerance to intravenous anesthetics in dogs. J. Pharmacol. Exp. Ther. 128: 192–200, 1960.

7. LeBlanc, A.E., Kalant, H. and Gibbins, R.J. Acute tolerance to ethanol in the rat. Psychopharmacologia 41: 43–46, 1975.

8. Chen, C.-S. A study of the alcohol-tolerance effect and an introduction of a new behavioural technique. Psychopharmacologia 12: 433–440, 1968.

9. LeBlanc, A.E., Gibbins, R.J. and Kalant, H. Behavioral augmentation of tolerance to ethanol in the rat. Psychopharmacologia 30: 117–122, 1973.

10. LeBlanc, A.E., Kalant, H. and Gibbins, R.J. Acquisition and loss of behaviorally augmented tolerance to ethanol in the rat. Psychopharmacology 48: 153–158, 1976.

11. Traynor, M.E., Schlapfer, W.T., Woodson, P.B.J. and Barondes, S.H. Tolerance to a specific synaptic effect of ethanol in *Aplysia*. *In* Biochemistry and Pharmacology of Ethanol, Vol. 2, E. Majchrowicz and E.P. Noble, eds. pp. 269–279. Plenum Press, New York, 1979.

12. Okamoto, M., Boisse, N.R., Rosenberg, H.C. and Rosen, R. Characteristics of functional tolerance during barbiturate physical dependency production. J. Pharmacol. Exp. Ther. 208: 906–915, 1978.

13. Goodwin, D.W., Powell, B., Bremer, D., Hoine, H. and Stern, J. Alcohol and recall: state-dependent effects in man. Science 163: 1358–1360, 1969.

14. Overton, D.A. State-dependent learning produced by depressant and atropine-like drugs. Psychopharmacologia 10: 6–31, 1966.

15. Siegel, S. Morphine analgesia tolerance: its situation specificity supports a Pavlovian conditioning model. Science 193: 323–325, 1976.

16. Sato, C., Matsuda, Y. and Lieber, C.S. Increased hepatotoxicity of acetaminophen after chronic ethanol consumption in the rat. Gastroenterology 80: 140–148, 1981.

17. Leeb-Lundberg, F., Snowman, A. and Olsen, R.W. Barbiturate receptor sites are coupled to benzodiazepine receptors. Proc. Nat. Acad. Sci. 77: 7468–7472, 1980.

18. Rottenberg, H., Waring, A. and Rubin, E. Tolerance and cross-tolerance in chronic alcoholics: reduced membrane binding of ethanol and other drugs. Science 213: 583–585, 1981.

Review

Kalant, H., LeBlanc, A.E. and Gibbins, R.J. Tolerance to, and dependence on, some non-opiate psychotropic drugs. Pharmacol. Rev. 23: 135–191, 1971.

8. Physical Dependence

Physical dependence is a straightforward pharmacological effect of ethanol itself, even though the abnormality is observable only after removal of the drug. There is an actual physical illness, one that turns out to be fairly easy to reproduce in laboratory animals. In this chapter, we will consider first some questions that arise as a result of clinical observation of the alcohol abstinence syndrome. These questions suggest experiments for the more controlled setting of the laboratory or the research ward. Many such experiments have been done in the relatively few years since we learned that laboratory animals could be made physically dependent on ethanol. The experimental results sometimes force a rethinking of clinical assumptions.

The clinical syndrome

Patients admitted to the hospital in a severely intoxicated condition often undergo a specific and dangerous episode—the alcohol withdrawal reaction or abstinence syndrome. The clinical picture has never been better described than in the 1953 study by Victor and Adams [1]. The Boston City Hospital, where they worked, is a large metropolitan hospital; it receives many alcoholics. (There were 266 patients admitted to the hospital in this two-month study. In addition, the authors estimated that ten times that many intoxicated patients were treated briefly in the emergency room—on the average, nearly two an hour.) The clinical picture described by Victor and Adams was complicated by disorders other than intoxication, including trauma, infection, malnutrition, and liver disease, which made it difficult to characterize the effects of ethanol per se. Nevertheless, they were able to describe the main features of the syndrome, seen with varying severity in most of their patients. Central and autonomic hyperactivity is the underlying abnormality, revealed by a family of related signs and symptoms.

A remarkable feature of the withdrawal reaction is its temporal pattern, which was first described in the Victor and Adams study and which has not yet been explained, or even reproduced in animal models. A tremulous phase ("the shakes") is seen on admission and indeed is often noted by the patients in the mornings during a period of sustained drinking. This can be severe enough to prevent the patient from feeding himself. During the first 24 hours, hallucinations are often experienced that are relatively benign without loss of orientation. About a day after cessation of drinking, convulsions may occur in severely affected individuals. These are identical to the tonic-clonic seizures of grand mal epilepsy, with loss of consciousness. Victor and Adams were not sure whether the convulsions occurred only in epileptic individuals or perhaps in "latent epileptics." It is now clear that withdrawal convulsions occur in patients who never have seizures on other occasions, and we therefore no longer think of these seizures as a manifestation of epilepsy but rather as a predictable component of a severe alcohol withdrawal reaction.

If the preceding binge has been prolonged or intense, the condition of these patients steadily worsens after withdrawal, and on the third or fourth day they may become confused and disoriented, with much more serious hallucinations. This is the onset of delirium tremens. There is a severe autonomic overactivity, with sweating, nausea, vomiting, diarrhea, and fever. The patient is agitated, picking at the bedclothes and shouting warnings or instructions to bystanders both real and imaginary. He or she is totally disoriented with respect to time and place. There may be a very high fever, in which case the prognosis is poor. The mortality of delirium tremens is difficult to estimate because of the concurrent infections, trauma, etc. With modern treatment, the mortality has been much reduced, but this remains a life-threatening condition.

These signs and symptoms present themselves in sequence. Patients who have both seizures and delirium almost invariably have the seizures first. The most severe signs are the latest to appear, and they occur only in the patients who have the most intense overall withdrawal reactions.

Experimental production of physical dependence

The rum fits experiment. Does drinking itself cause the syndrome described
by Victor and Adams, or is it caused by termination of drinking? Clinical
views had been marshalled on both sides of this question for decades.
Withdrawal reactions do not necessarily follow drinking bouts; most inebri-
ates simply sober up, without any such episode. Further, heavy drinkers
may have tremor or even mild hallucinations when they are still drinking.
The question was difficult to answer on the wards, but it was attacked
experimentally in 1955 in the famous "rum fits" study of Isbell et al., done
at the federal prison-hospital for narcotic addicts at Lexington, Kentucky
[2]. In that institution, important clinical pharmacological research had been
carried out over the previous 20 years, using the prisoner ex-addicts as
subjects. There had been experiments in which barbiturates were given
several times a day, and a mild withdrawal reaction was seen when they
were discontinued. A parallel experiment was planned using alcohol and
incorporating the principle learned from experiments on other addictive
drugs, namely that physical dependence follows a period of uninterrupted
intoxication. The subjects for the alcohol experiment were ten former morphine addicts, of whom six had histories of excessive use of alcohol. They
were observed very carefully indeed. At the start and repeatedly throughout the experiment, they were given complete physical examinations, blood
chemistry tests, EEGs, and psychiatric tests. Their nutrition was well controlled, with a 4000-calorie diet (plus snacks) and vitamin supplements added
near the end; they gained weight. Ethanol was given orally every few hours
around the clock, in doses sufficient to maintain the maximum degree of
intoxication that could safely be managed on the ward. The ten subjects
stayed on this regimen for varying periods of time, from a week to three
months. They became disheveled and quarrelsome, but did not appear to
be in bad health. Their alcohol intake was high even at the start, in relation
to the assumed amounts that a person can metabolize. They took over 10 g
per hour at first, and later increased that by about 30%, maintaining stable
blood levels on a high intake. The blood alcohol levels, monitored in three
patients, were generally elevated but showed a transient fall at two or three
weeks, with an accompanying decrease in intoxication ratings. This may
have marked the onset of metabolic tolerance. With a slight increase in
dose, they returned to their previous high blood alcohol concentration and
intoxicated state. Functional tolerance was manifested by the changed relationship between behavioral impairment and blood alcohol levels.

When these patients stopped drinking, they underwent a syndrome re-markably similar to that described by Victor and Adams. Most strikingly, the severity of the syndrome was related to the duration of drinking, as Table 8-1 shows. The patients had tremor and a variety of autonomic signs, as well as the hallucinations, convulsions, and delirium that we now know to be characteristic of the alcohol withdrawal syndrome. Tremor began with even a slight reduction in blood alcohol. Two patients had convulsions and one went into status epilepticus. Two were treated with barbiturates to alleviate the unexpectedly severe withdrawal reaction. All the patients re-covered after a few days and were fully back to normal at a three-month follow-up.

This important experiment showed definitively that rum fits and delir-ium tremens are alcohol withdrawal reactions. Convulsions and delirium were never seen during drinking. Furthermore, malnutrition and epilepsy were ruled out as causative agents. Since then there has never been any serious doubt that termination of alcohol intake can cause the severe syn-drome described by Victor and Adams.

Animal models. Thus, the first experimental alcohol withdrawal syndrome was actually produced in humans, and the animal models followed. For a

Table 8-1. Signs and symptoms after withdrawal of alcohol[1]

Subj.	Days	T	W	P	H	N	V	D	A	I	HR	F	HL	DO	C	Dur.
M	7	1	1	0	0	0	0	0	1	0	0	0	0	0	0	1
R	16	1	1	1	0	1	0	0	1	0	0	0	0	0	0	2
B	16	1	1	1	0	0	0	0	0	0	0	0	0	0	0	3
TM	34	2	2	2	2	1	0	0	1	0	0	0	0	0	0	3
JR	48	3	3	3	2	2	2	1	2	2	3	1	2	0	1	5
TY	48	4	4	4	4	3	3	3	2	4	3	3	4	4	0	?
S	55	4	3	3	2	2	2	1	1	2	2	1	0	0	0	5
C	78	4	4	4	4	4	4	4	2	?	4	4	?	4	7	?
JK	78	4	4	3	3	2	2	1	2	4	3	3	4	1	0	8
A	87	3	3	3	1	2	2	1	2	2	3	1	2	0	0	8

1. Scores of 0 to 4 were assigned to increasing severity of each sign, except convulsions, where the numeral indicates the number of convulsions. Question marks indicate uncertainty because the withdrawal reaction was terminated by administration of barbiturates. Days = duration of drinking period; Dur. = duration of withdrawal reaction, in days. From Isbell et al. [2].

T	= tremor	V	= vomiting	F	= fever
W	= weakness	D	= diarrhea	HL	= hallucinations
P	= perspiration	A	= anorexia		(visual and auditory)
H	= hypertension	I	= insomnia	DO	= disorientation
N	= nausea	HR	= hyperreflexia	C	= convulsions

long time, attempts to make animals physically dependent on alcohol had failed. Inadequate alcohol intake was actually the cause, but it was originally thought that there was some mystical property of the human mind that made only man capable of becoming dependent on alcohol. The first indication of physical dependence in animals was seen in monkeys. Several investigators reported seeing withdrawal hyperexcitability following periods of spontaneous abstinence interspersed among "binges" in monkeys that were self-injecting ethanol, but these workers did not describe the withdrawal reaction in detail. Essig and Lam [3] did the first convincing experiment on animal withdrawal reactions in 1968, when they repeated Isbell's experiment with dogs, again using the principle of continuous intoxication. The ethanol was given through indwelling gastric cannulae, as often as every four hours during the last two weeks of the eight-week experiment. No seizures or hallucinatory behavior occurred during the drinking period, but on withdrawal a syndrome appeared that was obviously analogous to the human abstinence syndrome. The dogs showed tremor, hyperreflexia, spasticity, and one or more convulsions. Two dogs died soon after having a convulsion; in all, four of the eight dogs died within 48 hours after withdrawal. One dog was observed to behave as though it were having hallucinations, turning its head as if looking at a moving object and snapping at objects unseen by the observer. Other workers have documented such hallucinatory behavior on film, both in dogs and in monkeys. This was the first convincing animal model for the alcohol withdrawal reaction.

Several other models followed promptly. The next year, Freund [4] reported the production of alcohol physical dependence in mice. He administered ethanol in a simple liquid diet (a commercially available human weight-control potion). Many previous attempts to make mice or rats dependent on ethanol had failed when the alcohol was simply added to the drinking water, even when the ethanol solution was the only available fluid. The failure was probably due to the fact that rodents drink mostly at night and they metabolize ethanol rapidly; thus, they are sober most of each day. The liquid diet (for some reason) was taken more evenly around the clock, so the mice stayed intoxicated. When the alcohol was removed from the diet after only four or five days of intoxication, a withdrawal reaction was observed. Semiquantitative ratings of its intensity were assigned on the basis of tremor, abnormal movements, convulsions, and death. This was the first of many rodent models of alcohol dependence. Since then, alcohol physical dependence has been evoked in monkeys, dogs, cats, rats, and mice. The necessary trick is to keep the animals intoxicated almost all the time; this requires some special procedures in rodents, since they metabolize alcohol rapidly.

Physical dependence as an adaptive response. The next question that arises from the clinical material is one of interpretation. How could the discontinuing of a drug possibly cause such a violent reaction? As explained in Chapter 7, many researchers now believe that the addiction process is basically an expression of adaptation. Withdrawal reactions may be the result of returning to the original drug-free state so rapidly that deadaptation cannot keep pace. Then the organism would find itself adapted to a condition that no longer exists and would show abnormalities in the opposite direction to those of the initial drug effect, i.e., CNS excitation would replace sedation. Some hypothetical target systems have been proposed. For example, it has been suggested that redundant neuronal pathways might develop during intoxication, taking the place of neurons inhibited by ethanol. On withdrawal, both the original and the recently developed systems would be active, leading to the hyperexcitability. Models involving proliferation of receptors or induction of enzymes also have been worked out, but no one has found the actual mechanism.

A major difficulty with the adaptation hypothesis is that it does not explain the late components of the withdrawal reaction. According to this concept, the withdrawal syndrome should peak just when the drug disappears from its target site, because at this time there is the maximum imbalance between the adapted target system and the drug concentration. At later times, the target should return to normal and the syndrome should subside. Instead, it continues to worsen, even in the animal models. We cannot explain the late appearance of delirium tremens by the homeostat notion.

How much drinking does it take?

Alcoholics often turn up in the hospital in their forties or fifties, with a long history of heavy drinking. But Isbell's study showed that severe withdrawal reactions could develop in people with no significant alcohol history, after only a few weeks of drinking. Physical dependence may develop faster than alcoholism. Perhaps it takes many years for a social drinker to develop the ability to drink heavily over several days and thus to incur severe physical dependence. Only then will a brief, intense binge precede delirium tremens.

According to the homeostat hypothesis, the adaptive process should begin immediately on application of the stimulus (drug). Experimental evidence supports this prediction. In 1958, McQuarrie and Fingl [5] reported that a state of CNS hyperexcitability could be detected in mice after a single dose of ethanol. They measured the threshold for seizures evoked by pentylenetetrazol or electroshock. During the intoxication, the seizure threshold was increased because of the anticonvulsant action of ethanol, but a few hours later the threshold fell below that of the controls, indicating a state of seizure susceptibility (Fig. 8-1). This was actually the first demonstration of an alcohol withdrawal reaction in laboratory animals. Its importance was recognized only later, perhaps because the paper described only seizure thresholds, rather than a whole syndrome analogous to the human withdrawal reaction. The miniwithdrawal reaction after a single dose of ethanol suggests that withdrawal symptoms may be a part of the hangover malaise.

Plateau principle equations. Physical dependence, then, begins with the first drink and develops progressively over a period of weeks, if drinking

Fig. 8-1. *Withdrawal hyperexcitability after a single dose of ethanol in mice.* At different times after an oral dose of 4 g/kg, the seizure threshold for convulsions elicited by pentylenetetrazol was measured in different groups of mice. The ordinate shows the threshold as a multiple of the control threshold. The high ratio at 4 hours demonstrates the anticonvulsant effect of ethanol, and the low ratio at 8 hours shows that a state of hyperexcitability existed after the ethanol had been eliminated. Bars show the 95% confidence limits of the ratios. From McQuarrie and Fingl [5].

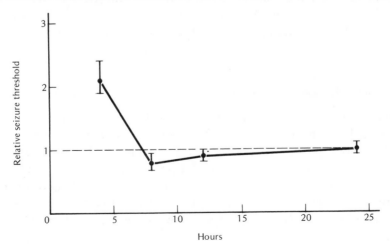

continues. In Isbell's experiment, withdrawal reactions were less severe after brief periods of drinking than after several weeks. More quantitative studies of the time course can be conducted with animals. For example, mice exposed continuously to ethanol show progressively severer with-drawal convulsions as the duration of intoxication is increased to one or two weeks; thereafter the withdrawal scores level off [6]. Such a time course can be expressed mathematically using equations developed under the ru-bric of the "plateau principle" [7]. The equations deal with a common sit-uation, applicable in many different pharmacological contexts. They de-scribe what happens when some substance is being formed (or taken up) at a constant rate and simultaneously being degraded (or eliminated) with first-order kinetics. The rate of change in the concentration of the substance, X, is described by

$$\frac{dX}{dt} = S - kX$$

where S is the constant rate of synthesis (in units of concentration per unit time), and k is the first-order decay constant. At the steady state, where $X = X_{ss}$,

$$\frac{dX}{dt} = 0$$

$$S = kX_{ss}$$

$$X_{ss} = \frac{S}{k}$$

In changing conditions, the level of X will follow any alteration in the rate of its synthesis or degradation. If the rate of synthesis is abruptly increased, X will begin to rise, but the higher it goes the faster is its breakdown, and a steady state develops again. The equations predict the magnitude and time course of change in X after a shift in rate of synthesis or breakdown. Integrating the original equation, we can solve for X at any time or for the halftime of a shift to the new steady state.

$$\int_{X_0}^{X} \frac{dX}{S - kX} = \int_{t_0}^{t} dt$$

or

$$\ln \frac{S - kX}{S - kX_0} = -kt$$

The rest of the derivation will be given here for the simple case where $X_o = 0$, i.e., where the initial synthesis rate is zero. The equations can easily be worked out by the interested reader for the case where X_o and S_o are not equal to zero. Solving for X from the above equation,

$$X = \frac{S}{k}(1 - e^{-kt})$$

$$X = X_{ss}(1 - e^{-kt})$$

And one can solve for the halftime of the shift to the new steady state, where

$$X = \frac{1}{2}X_{ss} \text{ and } t_{\frac{1}{2}} \text{ is the halftime}$$

$$\frac{1}{2} = 1 - e^{-kt_{\frac{1}{2}}}$$

The equation reduces to

$$t_{\frac{1}{2}} = \frac{\ln 2}{k}$$

Note that S has vanished; the time course of the shift depends only on the rate of breakdown of X, not on its present or former rate of synthesis.

Now let us apply the model to our postulated adaptive response to chronic ethanol intake. X is now a substance at the biochemical site of action of ethanol. New conditions, imposed by the presence of the drug, cause a shift to a new steady state level of X. According to the equations, the adaptation process is related to the turnover rate of the target system. If we are dealing with a protein, then the shift to the new steady state should have a halftime of hours or a few days, in accord with typical protein turnover rates.

A mouse model. To test these concepts in the laboratory, we must control the stimulus for adaptation and keep it constant. The stimulus is the presence of ethanol in the brain. In one such model [8], ethanol is administered by inhalation; mice are housed for a few days in a box through which a low concentration of ethanol vapor is kept flowing. One can stabilize the blood alcohol levels by treating the mice daily with a small dose of pyrazole, which inhibits alcohol dehydrogenase and thus retards the elimination of ethanol (Chapter 1). Its usefulness in this system depends on the fact that the inhibition is competitive with ethanol. Whenever the blood alcohol concentration rises, it partially overcomes the inhibition of its metabolism, so there is good feedback control and nicely stable concentrations are maintained for days. On withdrawal, after three days of such constant intoxication, at alcohol blood levels of 2 mg/ml, a reproducible syndrome of CNS hyperexcitability ensues that can be numerically evaluated by scoring the severity of convulsions elicited by handling the mice. The withdrawal scores rise for about 10 hours and then return to zero (Fig. 8-2). The peak heights of such curves or areas under them are reliable measures of the severity of withdrawal reactions. They are related to the alcohol exposure, that is, to the product of the blood alcohol concentrations and duration of alcohol

Fig. 8-2. *Withdrawal reaction scores in a mouse model of physical dependence.* After three days of mild but continuous intoxication produced by inhalation of ethanol, the mice were removed from the vapor chamber and scored hourly for convulsions on handling. The rise and fall of the scores illustrate the intensity and time course of the withdrawal reaction. The data are from several experiments using a total of 95 mice; bars show SEM. From Goldstein [9].

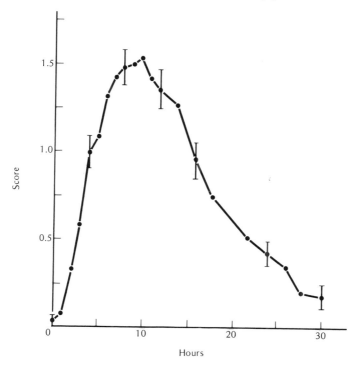

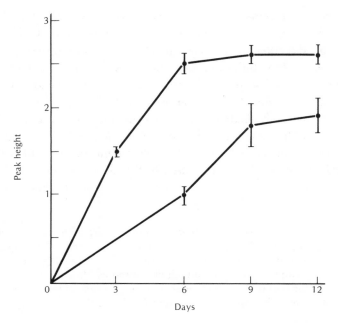

Fig. 8-3. *Rate of onset of physical dependence in mice.* Peak heights of curves like that shown in Figure 8-2 were used as measures of relative magnitude of physical dependence in groups of mice exposed to ethanol by inhalation for 3–12 days. The upper curve represents mice that had maintained blood ethanol concentrations of about 2 mg/ml; the lower curve, 1 mg/ml. Vertical bars show SEM for 9–95 mice per point. From Goldstein [6].

administration. These scores show the rate of onset of physical dependence, and we can compare it with the predictions of the homeostat hypothesis. Figure 8-3 shows the peak heights of withdrawal reactions of mice exposed for different lengths of time to either of two different blood alcohol concentrations. The scores rise to an apparent plateau, the height of which is related to the blood alcohol concentration. The time course is the same for the two curves; the halftime is a few days. According to our model, the time course reflects the turnover rate of the adapting system and does not depend on the drug dose. It should be the same for both curves, which it does seem to be. Further, the level of the plateau should be related to the intensity of the stimulus for adaptation, i.e., the drug concentration, and this also is the experimental result [6].

A model commonly used for rats is that of Majchrowicz [10]. He administered ethanol by gastric intubation for a period of four days. Stability of blood ethanol levels is achieved by giving the alcohol in divided doses, three to five times a day, and varying the size of the dose according to the behavioral state of the animal. Rats that are comatose skip a dose; those that are ataxic receive small doses, etc. The resultant withdrawal reaction includes hyperactivity, audiogenic seizures, spontaneous convulsions, and sometimes death. The syndrome peaks at 14–16 hours after withdrawal, much later than in mice where the maximum intensity is seen at about six hours if pyrazole is not used. (Pyrazole retards alcohol elimination and delays the development of the withdrawal reaction.) In both species, a severe withdrawal reaction is brought on by only a few days of ethanol administration, provided elevated blood levels are continuously maintained.

How then to explain the much slower but still progressive development of dependence among alcoholics and in Isbell's experiment? We can postulate that there is a process of simultaneous onset and decay of the dependent state and that whenever blood alcohol levels fall, as they will do with intermittent alcohol intake, deadaptation begins. When blood levels fluctuate, there is an alternating wax and wane of the magnitude of the dependence. Is the decay of dependence fast enough to matter, over periods of hours when mice or people sober up between drinks? To examine this experimentally, we looked for carry-over of physical dependence in mice from one cycle of intoxication to the next, a cycle being a standard three-day alcohol exposure [6]. In different experiments we allowed intervals of different lengths between the intoxication periods. We expected that withdrawal scores would cumulate from cycle to cycle only if the interval between cycles was short enough to prevent the complete decay of the previously acquired dependent state. We found such a cumulation only when the interval between cycles was very short indeed—12 hours or less. When there was a day or more between cycles, the withdrawal scores remained the same, cycle after cycle, indicating that the dependence had completely disappeared during the interval. The halftime for decay of dependence must be a few hours, which probably explains why it had been so difficult to make mice dependent by simply putting alcohol in the drinking water. If they did get drunk during their nighttime drinking period, they would quickly sober up, and during the day the small amount of accrued dependence would disappear. It is not known whether there can be any transfer of this information to human drinkers. Continuous drinking may be much more dangerous than intermittent drinking, and an occasional day skipped might have a beneficial effect.

These experiments are based on a single withdrawal sign, a specific type of convulsion. Since the withdrawal reaction has many components with different time courses, several mechanisms may be simultaneously at work, and the rapid decay of physical dependence may not apply to all its aspects. Some components of the dependent state with slower time course may account for carry-over of dependence from one drinking episode to the next, as is sometimes observed, both experimentally and clinically. Quantitative data on this point are few.

An experiment by Branchey et al. [11] is the best evidence for carry-over in animals. They treated two groups of rats with ethanol in a liquid diet for four days and measured withdrawal reactions. One group had had a course of prior alcohol administration, ending two weeks earlier; the other group had not. In the four-day test when both groups received alcohol diets, the group that had been previously treated had the severer withdrawal syndrome. The two groups drank the same amount of alcohol during the four-day intoxication period, which is of course crucial. Gross and co-workers [12] have assembled evidence for carry-over of tolerance and physical dependence from one binge to another in human alcoholics.

Drug treatment of withdrawal reactions

Treatment of the human abstinence syndrome is a pharmacologically rational procedure. Clearly the syndrome could be suppressed by ethanol itself. Indeed, this drug is the almost universal choice of alcoholics faced with morning shakes. Ethanol, however, is not suitable for medical treatment of withdrawal reactions because it is too rapidly and too variably eliminated. One wants a stable blood level of a sedative drug that can substitute for ethanol, either at its actual site of action or at a site that evokes comparable behavioral (CNS depressant) effects. Barbiturates will suppress alcohol withdrawal reactions, including the convulsions. Other anticonvulsants such as phenytoin are seldom indicated, since convulsions are relatively rare and unpredictable. A more general sedative action is needed. Ethanol surrogates such as paraldehyde and chloral hydrate have been widely used for controlling the withdrawal hyperexcitability. They have a reputation for safety that derives mainly from restricting their use to hospital settings. Paraldehyde, however, can be dangerous because of its tendency to break down into acetic acid if improperly stored. Despite the obnoxious odor of paraldehyde, some alcoholics switch their addiction from ethanol to paraldehyde. For a time, phenothiazines were used to treat alcohol withdrawal reactions, presumably because of their antischizophrenic actions, which seemed appropriate for the withdrawal hallucinatory state. However, it is now clear that phenothiazines have a tendency to increase seizure susceptibility. Results of several well-controlled clinical trials suggest that withdrawal reactions are worsened by administration of chlorpromazine or promazine [13].

Benzodiazepines are the drugs of choice for treatment of alcohol withdrawal reactions, replacing the relatively dangerous barbiturates and paraldehyde. Chlordiazepoxide and diazepam are long-lasting sedative drugs, partly because they have active metabolites that are slowly eliminated. They are ethanol surrogates, they last through a substantial part of the withdrawal period, and they are safe. It makes sense to treat this condition with a drug that can safely be given over a period of days with the blood levels slowly declining at about the same rate that recovery from the dependent state occurs. The sedative, antianxiety, and anticonvulsant properties of the benzodiazepines are useful in this condition. Nevertheless, even with benzodiazepines it is not always possible to suppress delirium tremens once it has begun.

Some patients may transfer their addiction from alcohol to benzodiazepines but this is a minor problem (compared to alcoholism) because benzodiazepines are much less toxic than ethanol and they produce mild withdrawal reactions if any.

Summary

Despite centuries of clinical observation, it has not always been recognized that rum fits and delirium tremens are the result of withdrawal of alcohol. The question was settled definitively by an analytical clinical study and a controlled experiment with human subjects. Animal models have since been used to examine the time course and dose relations of the dependent state. Physical dependence can develop rapidly: some withdrawal signs are seen after a single dose of ethanol. A mathematical model, based on the plateau principle, describes the progress of physical dependence as an adaptive response. Carry-over of physical dependence from one binge to the next is seen in some experiments or clinical situations but not in others. No animal models reproduce or explain the curious sequence of increasingly severe signs that occurs in the human abstinence syndrome. Controlled clinical trials have established benzodiazepines as the drugs of choice in treating alcohol withdrawal reactions.

References

1. Victor, M. and Adams, R.D. The effect of alcohol on the nervous system. Res. Publ. Assoc. Res. Nerv. Mental Dis. 32: 526–573, 1953.
2. Isbell, H., Fraser, H.F., Wikler, A., Belleville, R.E. and Eisenman, A.J. An experimental study of the etiology of "rum fits" and delirium tremens. Quart. J. Stud. Alc. 16: 1–33, 1955.
3. Essig, C.F. and Lam, R.C. Convulsions and hallucinatory behavior following alcohol withdrawal in the dog. Arch. Neurol. 18:626–632, 1968.
4. Freund, G. Alcohol withdrawal syndrome in mice. Arch. Neurol. 21: 315–320, 1969.
5. McQuarrie, D.G. and Fingl, E. Effects of single doses and chronic administration of ethanol on experimental seizures in mice. J. Pharmacol. Exp. Ther. 124: 264–271, 1958.
6. Goldstein, D.B. Rates of onset and decay of alcohol physical dependence in mice. J. Pharmacol. Exp. Ther. 190: 377–383, 1974.
7. Goldstein, A., Aronow, L. and Kalman, S.M. Principles of Drug Action. The Basis of Pharmacology. Second edition, p. 311. John Wiley and Sons, New York, 1974.
8. Goldstein, D.B. Relationship of alcohol dose to intensity of withdrawal signs in mice. J. Pharmacol. Exp. Ther. 180: 203–215, 1972.
9. Goldstein, D.B. Alcohol withdrawal reactions in mice: effects of drugs that modify neurotransmission. J. Pharmacol. Exp. Ther. 186: 1–9, 1973.
10. Majchrowicz, E. Induction of physical dependence upon ethanol and the associated behavioral changes in rats. Psychopharmacologia 43: 245–254, 1975.
11. Branchey, M., Rauscher, G. and Kissin, B. Modifications in the response to alcohol following the establishment of physical dependence. Psychopharmacologia 22: 314–322, 1971.
12. Gross, M.M. and Best, S. Behavioral concomitants of the relationship between baseline slow wave sleep and carry-over of tolerance and dependence in alcoholics. Adv. Exp. Med. Biol. 59: 633–643, 1975.
13. Kaim, S.C., Klett, C.J. and Rothfeld, B. Treatment of the acute alcohol withdrawal state: a comparison of four drugs. Am. J. Psychiat. 125: 1640–1646, 1969.

Review

Goldstein, D.B. Animal studies of alcohol withdrawal reactions. *In* Research Advances in Alcohol and Drug Problems, Israel, Y., Glaser, F.B., Kalant, H., Popham, R.E., Schmidt, W. and Smart, R.G., eds., Vol. 4. Plenum, New York, 1978, pp. 77–109.

9. Voluntary Intake of Ethanol by Laboratory Animals

Many investigators have undertaken experiments designed to induce laboratory animals voluntarily to consume alcohol in quantities sufficient to injure their health or cause them to become addicted. Sometimes the goal is to develop a complete animal model of alcoholism. Criteria for such a model have been defined; they usually include voluntary intake of ethanol by the oral route, in amounts that produce tolerance and physical dependence, with relapse after periods of abstinence, etc. It is my view that this is a fruitless endeavor at present. Alcoholism is an enormously complex disease, with facets that are biochemical, medical, psychiatric, psychological, and more. A model of alcoholism that included all the variables would be as complicated as the actual alcoholic patient, and just as hard to understand. It seems to me more profitable to use an analytical approach, separating the variables and studying one at a time, dissecting out parts of the syndrome that we can hope to comprehend in a controlled study. It is feasible to direct an investigation specifically at factors which control the amount of ethanol that animals (or people) will voluntarily consume. These studies, properly controlled, are important for the experimental investigation of alcoholism. This should be a gratifying field of work since positive results in either direction are exciting; conditions that lead animals to increase their voluntary consumption of ethanol may represent a model for human drinking, whereas decreased intake could be a model for treatment.

Unfortunately, much of the literature in this field is conflicting or indecisive. Factors such as stress, nutrition, and endocrine agents may well influence voluntary ethanol consumption, but the experiments so far have not shown clear or reproducible effects. These studies have been thoroughly reviewed by Myers and Veale [1] and will not be covered here.

Alcohol solutions as sole source of drinking fluid

Alcohol intake is truly voluntary only if the animals lose nothing when they choose not to partake of it. By contrast, when an ethanol solution is the sole source of drinking fluid, the intake cannot be said to be voluntary, even though it is taken by the animals themselves, rather than administered forcibly. Richter showed, as long ago as the 1920s, that rats could thrive on ethanol solutions as their only fluid, and that they survived on ethanol concentrations as high as 24% [2]. The animals evidently obtained enough fluid without exceeding the rate at which they could metabolize ethanol. They did not become intoxicated nor did they show any withdrawal reactions when the ethanol was discontinued. Many experimenters have simply offered ethanol in the drinking fluid of experimental animals and looked for pharmacological effects. Seldom have these experiments produced useful findings with respect to physical dependence, increased voluntary alcohol intake, or other signs relevant to alcoholism. Rats find ways to defeat the purpose of a voluntary intake experiment. For example, when offered a 40% ethanol solution as their only source of drinking fluid, they learned to drink so slowly that the drop hanging from the drinking tube lost most of its ethanol by evaporation between sips [3].

Schedule-induced polydipsia. One rather bizarre technique makes rats drink more ethanol than they would ordinarily choose. Falk and co-workers have found that certain schedules of food reinforcement caused rats to drink enormous amounts of water during operant conditioning sessions. This "schedule-induced polydipsia" required no lever-pressing activity, only delivery of food in small portions at frequent intervals. Rats took enough ethanol to produce physical dependence when ethanol (5%) was substituted for drinking water during six polydipsia sessions daily [4]. Blood ethanol concentration remained elevated throughout the day and night for a period of three months. Discontinuation of ethanol administration was followed by a withdrawal reaction, demonstrated by audiogenic seizures. Schedule-induced polydipsia is poorly understood, but has proved reproducible and has been used for several studies of physical dependence.

Self-administration of ethanol by monkeys

Techniques for intravenous self-administration of drugs by monkeys or rats, developed at the University of Michigan, have been used extensively for study of the reinforcing property of drugs, which is related to their addiction liability. Unrestrained animals are provided with an indwelling intravenous catheter and can press a lever to receive a drug injection. Among the behavioral variables to be studied are initiation and maintenance of drug self-administration and the temporal pattern of drug intake. The relative potency of drugs as reinforcers can be estimated by observing how hard the animals will work to obtain an injection. The schedule of reinforcement (drug administration) can be set to any desired fixed ratio. That is, the injection could be obtained with each lever-press or every tenth or hundredth, etc. The ratio can be progressively increased to see how much the animals will tolerate before giving up. This type of experiment has shown, for example, that cocaine is extraordinarily potent, since animals will press levers hundreds of times for a cocaine injection. There is an excellent correspondence between the set of drugs that monkeys will self-inject and those that people become addicted to. This method has been much used in screening new drugs for their "addiction potential."

The Michigan group, in one of their earliest papers [5], reported that monkeys will self-inject ethanol, and subsequent research has confirmed their findings. Figure 9-1 shows some of their data. Monkeys can sustain a moderately high ethanol intake for a few months, and some animals will become physically dependent. They often drink for a couple of weeks and then stop, producing a withdrawal reaction. By restricting access to ethanol to periods of three hours daily, Winger and Woods [6] produced stable and reproducible behavior patterns that could be further manipulated and analyzed, but this procedure does not allow as high a daily intake as continuous access does.

Mello and Mendelson have shown that it is difficult to produce dependent monkeys by means of an oral self-administration paradigm [7]. Oral ethanol is clearly aversive. When monkeys are required to lick an ethanol dispenser in order to avoid a shock they use ingenious methods to postpone the shock without consuming ethanol. But by linking food presentation with ethanol availability in food-deprived monkeys, Meisch and co-workers [8] have been able to set up conditions where oral ethanol serves as a reinforcer. The animals will then press levers more readily for alcohol than for

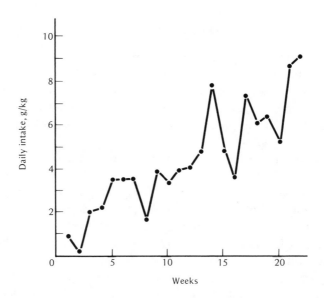

Fig. 9-1. *Intravenous self-administration of ethanol by a monkey.* The record shows the pattern of intake of a single animal self-injecting doses of 0.2 g/kg with each press of a lever. The monkey injected up to 40 doses a day, often producing signs of intoxication, and generally increased its intake over the 22-week period. However, the intake on individual days (not shown) revealed many brief interruptions, during which the monkey sobered up and showed withdrawal signs. From Deneau et al. [6].

water. High levels of blood alcohol have been reported, but physical dependence has not been studied with this model.

Alcohol preference model

The truly voluntary alcohol intake that many investigators seek is a situation where the animal has nothing to lose by refusing the ethanol. This can be done by offering a choice of drinking fluid, ethanol solution versus water, in two drinking bottles. The positions of the bottles are changed frequently to prevent the development of a position habit that might confound the data. The alcohol intake data can be expressed as the amount of absolute ethanol, in grams per kilogram of body weight per day or as alcohol "preference"—the ratio of volume of ethanol solution to the total fluid intake. Preference will be affected by the ethanol concentration and also by the animal's total fluid requirement. Rats decrease their preference ratio as the ethanol concentration is increased, but this tells us nothing about the total ethanol intake, in grams per kilogram body weight. Similarly, conditions that modify the total fluid consumption do not necessarily affect the absolute intake of ethanol.

A particular measure of preference has been used by Myers and coworkers for many years. They offer a choice of alcohol or water over a period of 12 days, during which time the ethanol concentration is progressively increased from 3% to 30%. Rats will generally accept the lower concentrations of ethanol, but they drink less and less ethanol, and more and more water, as the ethanol concentration is increased. This system can clearly be used to study effects of drugs and other procedures on voluntary alcohol intake. Its main disadvantage is that it necessarily brings time into the equation as an unassessed variable. During the 12-day test sequence, some tolerance may develop. It seems unlikely that a naive rat encountering a choice between water and 25% ethanol will take as much ethanol as a rat given this choice after it has been imbibing ethanol for several days. Thus, when it is reported that some drug or some stress, for example, shifts the preference curve one way or the other, we cannot be sure of the role of tolerance.

Myers has used his preference regimen to show the effects of various procedures on alcohol intake. Infusion of tiny doses of ethanol into the cerebral ventricles every 15 minutes prior to the 12-day preference sequence produced a dose-related increase in alcohol preference [9]. However, many other laboratories have examined voluntary alcohol intake after chronic ethanol administration and have found little or no effect. Myers reported that injection of p-chlorophenylalanine (pCPA) reduced preference for ethanol [10]. Because pCPA depletes brain serotonin by inhibiting the first step in its synthesis, the hydroxylation of tryptophan, it appeared that serotonin might be involved in alcohol intake—an interesting lead. However, another interpretation is more likely, namely that the reduced alcohol intake was the result of conditioned aversion. That is, the rats had associated the novel taste of ethanol with the dysphoria or illness caused by the pCPA. Nachman et al. [11] showed that rats treated with pCPA reduced their intake of saccharin, a normally desired solution. This is an example of "bait shyness," a form of conditioned aversion with a gustatory stimulus.

Garcia and co-workers have shown that conditioning with taste cues differs strikingly in time course from the better-known conditioning with visual, auditory, or olfactory cues [12]. Animals need the latter for escape from predators or for finding food, and reinforcement must follow immediately after the cue. By contrast, animals must be able to select nutritious foods and avoid toxic ones, even when the results of ingestion may not be evident for hours after intake. Taste must become associated with a much later result. Animals that survive an encounter with a poisoned bait will avoid that flavor thereafter. This is exactly what can happen in a preference test if the drug or procedure used to affect preference produces a toxic reaction in the animal. (It need not even be discernible, as conditioned aversion can develop in an anesthetized animal.) This is one of many pitfalls in research on alcohol voluntary intake. Spurious conclusions can be avoided by finding out whether the procedure changes the intake of another solution, as well as that of ethanol. No claim that a certain substance or situation reduces ethanol intake can be accepted without such controls.

The TIQ experiment. The chemistry and pharmacological activity of tetrahydroisoquinolines (TIQs) were described in Chapter 2. These compounds are condensation products of aldehydes and catecholamines. It has often been suggested that TIQs may mediate some chronic effects of ethanol, but experimental results have been disappointing. Recently, however, Myers and Melchior [13] have shown that several TIQs, infused intracerebroventricularly during the entire 12-day preference schedule described above, caused rats to drink much more ethanol than the controls. TIQs were infused in small volumes of artificial cerebrospinal fluid every 30 minutes around the clock. The ethanol intake of these animals was the largest yet reported for rats with any totally voluntary procedure. The intake of TIQ-treated rats was around 8 g/kg per day, on the average, but varied considerably. Strangely, there was no clear dose relation; the lowest TIQ dose sometimes caused the greatest increase in alcohol intake. Several TIQ derivatives produced this result, but unsubstituted tetrahydroisoquinoline itself was apparently much weaker. Controls infused with artificial cerebrospinal fluid showed no unusual alcohol preference. Active substances included tetrahydropapaveroline (the condensation product of dopamine and the aldehyde formed from dopamine by the action of monoamine oxidase), salsolinol (the dopamine-acetaldehyde condensation product), and a β-carboline (a condensation product of acetaldehyde with an indoleamine). Figure 9-2 shows the salsolinol data. The increased alcohol preference was apparently permanent, as the pattern of increased voluntary ethanol consumption was essentially unchanged on retest a month after the TIQ treatment.

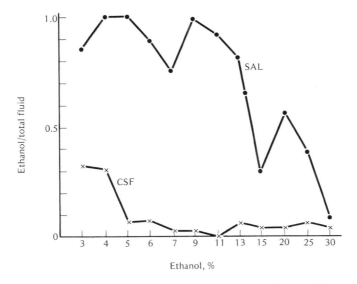

Fig. 9-2. *Effect of salsolinol on alcohol preference in Myers's paradigm.* The daily intake of ethanol is shown as percent of fluid intake, in a setup where the rats had a choice of drinking water or ethanol. The concentration of ethanol was increased daily from 3% to 30%, as shown on the abscissa. During this 12-day preference test, either artificial cerebrospinal fluid (CSF) (×) or salsolinol (SAL), 1 μg/μl (●), was injected into the cerebral ventricles in volumes of 0.4 μl every 30 minutes. The salsolinol rats (N = 4) took much more ethanol than did the CSF rats (N = 3) over the entire concentration range. The highest doses taken were about 7 g/kg of ethanol per day. From Myers and Melchior [14].

This intriguing experimental finding has been confirmed in another laboratory. Duncan and Deitrich [14] were able to replicate the essential features of the Melchior and Myers experiments. Rats infused with tetrahydropapaveroline or salsolinol did increase their alcohol intake in the 12-day preference test, and the increased preference persisted as long as ten months thereafter. Reasonable dose-responses were obtained with salsolinol and tetrahydropapaveroline, as shown in Figure 9-3, although the effect of tetrahydropapaveroline fell off at the highest dose. The maximal ethanol intake of individual rats, about 4 g/kg per day, produced only rather low blood alcohol concentrations.

The Duncan-Deitrich experiments only partially confirmed those of Melchior and Myers, and other investigators have failed to reproduce the results at all. This remains a highly controversial topic.

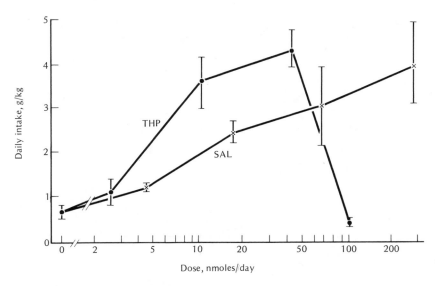

Fig. 9-3. *Dose-relation of alcohol intake produced by intraventricular injections of TIQs.* Each point represents a whole preference test (similar to those shown in Fig. 9-2) conducted with the indicated dose of THP or salsolinol (SAL). The alcohol intake is shown on the ordinate as the mean of daily intakes during the whole preference test, while the concentration of the ethanol in the drinking tube was raised from 3% to 30%. The daily dose of TIQ (abscissa) was administered in 48 injections into the cerebral ventricles. From Duncan and Deitrich [15].

Summary

Attempts to develop a complete animal model of alcoholism have not been successful, because it has proved difficult to find conditions under which animals will voluntarily drink themselves into a state of physical dependence. When ethanol solutions are the sole source of drinking fluid, the intake is usually insufficient to produce pharmacological effects. Certain schedules of food presentation can induce polydipsia. If ethanol is offered in such a situation, so as to spread the drinking evenly over each day, physical dependence can be produced. Monkeys will self-administer ethanol intravenously in amounts sufficient to produce intoxication and physical dependence, but oral self-administration is less effective, apparently because the taste of ethanol is aversive. Myers has been studying alcohol preference, i.e., ethanol intake in the presence of available water. Recently these experiments have shown that intracerebroventricular infusion of tetrahydroisoquinolines over a 12-day period produces a striking increase in alcohol intake which is apparently permanent.

References

1. Myers, R.D. and Veale, W.L. The determinants of alcohol preference in animals. In The Biology of Alcoholism, Kissin, B. and Begleiter, H., eds. Vol. 2, Plenum, New York, 1972, pp. 131–168.
2. Richter, C.P. Alcohol as a food. Quart. J. Stud. Alc. 1: 650–662, 1941.
3. Sohler, A., Burgio, P. and Pellerin, P. Changes in drinking behavior in rats in response to large doses of alcohol. Quart. J. Stud. Alc. 30: 161–164, 1969.
4. Falk, J.L., Samson, H.H. and Winger, G. Behavioral maintenance of high concentrations of blood ethanol and physical dependence in the rat. Science 177: 811–813, 1972.
5. Deneau, G., Yanagita, T. and Seevers, M.H. Self-administration of psychoactive substances by the monkey. A measure of psychological dependence. Psychopharmacologia 16: 30–48, 1969.
6. Winger, G.D. and Woods, J.H. The reinforcing property of ethanol in the rhesus monkey. I. Initiation, maintenance and termination of intravenous ethanol-reinforced responding. Ann. New York Acad. Sci. 215: 162–175, 1973.
7. Mello, N.K. and Mendelson, J.H. Factors affecting alcohol consumption in primates. Psychosom. Med. 28: 529–550, 1966.
8. Meisch, R.A. and Henningfield, J.E. Drinking of ethanol by rhesus monkeys: experimental strategies for establishing ethanol as a reinforcer. Adv. Exp. Med. Biol. 85B: 443–461, 1977.
9. Myers, R.D. Alcohol consumption in rats: effects of intracranial injections of ethanol. Science 142: 240–241, 1963.
10. Myers, R.D. and Veale, W.L. Alcohol preference in the rat: reduction following depletion of brain serotonin. Science 160: 1469–1471, 1968.
11. Nachman, M., Lester, D. and Le Magnen, J. Alcohol aversion in the rat: behavioral assessment of noxious drug effects. Science 168: 1244–1246, 1970.
12. Garcia, J., Hankins, W.G. and Rusiniak, K.W. Behavioral regulation of the milieu interne in man and rat. Science 185: 824–831, 1974.
13. Myers, R.D. and Melchior, C.M. Differential actions on voluntary alcohol intake of tetrahydroisoquinolines or a beta-carboline infused chronically in the ventricle of the rat. Pharmacol. Biochem. Behav. 7: 381–392, 1977.
14. Duncan, C. and Deitrich, R.A. A critical evaluation of tetrahydroisoquinoline induced ethanol preference in rats. Pharmacol. Biochem. Behav. 13: 265–281, 1980.

Review

Meisch, R.A. Ethanol self-administration: intrahuman studies. Adv. Behav. Pharmacol. 1: 35–84, 1977.

10. Genetics

Are there inherited differences among individuals that make some people more likely to become alcoholics than others? It is safe to assume that there are, since our genes affect every aspect of our lives. This is not to say that some people are helplessly born to alcoholism. The process of becoming an alcoholic has many components, which may include taste perception, sensitivity to acute intoxication, metabolic pathways, and personality factors such as sociability. All of these are influenced by nature as well as nurture, so we may expect from the start that both inheritance and environment will have some influence on vulnerability to alcoholism.

It is well known that some human societies have a much higher prevalence of alcoholism than others, suggesting a genetic component. But drinking patterns are so strongly controlled by cultural factors that the role of genetic differences between populations is difficult to discern. There is abundant evidence that alcoholism runs in families. We know that as many as a third of the first-degree male relatives (fathers, sons, brothers) of alcoholics are themselves alcoholics. But families share environment as well as genes, so it is quite possible that this is due to the chaotic situation in a household that includes an alcoholic, or to the family habit of turning to the bottle in every crisis. Let us look at the few attempts to answer the question more directly. They all give us the same message: there is something inherited about susceptibility to alcoholism.

Alcoholism in families

Twin studies. Genetic differences in human drinking behavior have been studied in three different twin studies comparing pairs of monozygotic (MZ) twins with dizygotic (DZ) pairs (identical versus fraternal twins). Differences between the two members of a MZ pair must be environmental, since their genes are identical; therefore, the *additional* difference between members of DZ pairs must be genetic. Whenever genetic factors are involved, there will be greater concordance within identical twin pairs than within fraternal twin pairs. The idea of parceling out the intrapair differences into only two components, genetic and environmental, ignores the very probable interactions between the two (for example, people who look alike may be treated alike by others), but it is a good way to start on a complex problem. A basic assumption of twin studies is that MZ pairs and DZ pairs share their environment to the same extent. This is known to be not quite true. MZ pairs, for example, tend to live together longer than DZ pairs do. This could lead to false positive results in any twin study, because some of the similarity ascribed to genetic identity in MZ pairs would in fact be due to environmental similarity.

One classic twin study on alcoholism was that of Partanen and co-workers in Finland [1]. As a sample, they used all the male twin pairs born in the 1920s in Finland and surviving in 1958 (aged 28–37). Zygosity was determined in the 902 pairs by physical measurements and in most cases by four to seven blood group tests. This left room for considerable error in determining zygosity. Methods are much better today, since many more antigens are available, and zygosity can be determined with little possibility of error. Such an error is always in the conservative direction, underestimating heritability, because some DZ pairs are taken to be MZ. (Any difference in physical measurement or blood groups will correctly assign the twin pair to the DZ group, but the absence of observed differences does not insure that they are MZ.) Each twin was separately questioned about drinking habits, including estimates of frequency of drinking episodes and the amount of alcohol taken each time, along with the duration of the episode in hours or days. The answers were grouped into a few factors, called "Density" (meaning frequency), "Amount" (per episode), and "Loss of Control" (the subject's own estimate, on a graded scale, of whether he could control his drinking). The latter was intended as a graded measure of alcoholism, as was a factor called "Social Consequences" (arrests for drunkenness, loss of job, etc.). For each factor, the intrapair variance was computed separately for the MZ and DZ pairs. An estimate of heritability of each behavior could then be computed from comparison of the variances of the two types of twin pairs. The equations are simple:

$$w^2 = \frac{1}{N}\sum_{ij}(x_{ij} - \bar{x}_i)^2$$

where w^2 is the intrapair variance and N is the number of pairs. Here x_{ij} is an individual score and x_i is the mean score of the ith pair.

Thus,

$$w^2 = \frac{1}{2N}\sum_i(x_{i1} - x_{i2})^2 = \frac{1}{2N}\sum_i \text{ (difference between twins in a pair)}^2$$

Then the heritability, H, is defined by

$$H = \frac{w^2(\text{DZ}) - w^2(\text{MZ})}{w^2(\text{DZ})}$$

H varies from zero to one. It is zero when the variance within MZ twin pairs is as great as that of DZ pairs, and it is 1.0 when the variance within MZ twin pairs is zero. For the factors called Density and Amount, the investigators found a statistically significant degree of heritability—a value around 0.5. But for Loss of Control and Social Consequences (indicating addiction), no significant genetic component was seen. This does not quite make sense, because one would expect some social consequences to go along with a genetically determined predisposition to drink a lot, but the study does suggest a substantial genetic contribution to drinking habits.

Another twin study was done in Sweden by Kaij [2]. The sample was smaller, but was enriched for alcohol abusers. Subjects were male twins in southern Sweden, one or both of whom was registered with the Temperance Boards. Swedish Temperance Boards are agents for social intervention in incipient alcoholism. Any kind of alcohol-related misbehavior may be reported to the local Board. Registry of an individual with the Temperance Board thus means that the person has had some sort of trouble with drink, but it does not identify him or her as an alcoholic. Kaij interviewed both twins in each pair, and added data from the Temperance Board records about arrests, hospitalizations, etc. The extent of alcohol consumption and its consequences was estimated on a scale of zero to four by combining all these data. The results were reported as the difference between members of each twin pair, taking MZ and DZ pairs separately; i.e., Kaij tabulated the number of pairs that were concordant or that had each degree of dis-

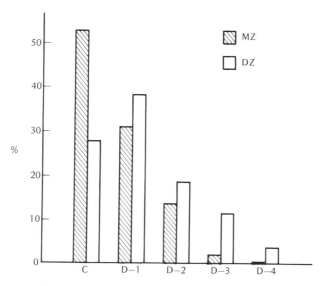

Fig. 10-1. *Concordance and discordance in drinking habits within twin pairs.* Concordance (C) means that both members of a pair had the same drinking-habits score. D-1 to D-4 designate pairs whose members differed by one to four points on the five-level scale, i.e., different degrees of discordance. The data show that monozygotic (MZ) pairs were often concordant and never strongly discordant, in contrast to dizygotic (DZ) pairs. From Kaij [2].

cordance and assessed the significance of the distribution by chi-square. Figure 10-1 shows the distribution of these differences; 54% of MZ twins and only 28% of DZ twins were concordant. Further, DZ twins were more heavily represented among the pairs with the greatest intrapair differences. For example, 11% of DZ pairs and only 2% of MZ pairs differed by three classes or more. A shortcoming of Kaij's study is that he did all the interviews himself, with full knowledge of the Temperance Board reports of the subject or his brother. This may well have biased the results, but we have no way of assessing the magnitude of error.

Half-sibs and adoptees. Another way to go about it, avoiding the assumptions of twin studies, is to examine subjects who have been raised apart from their biological parents. Two such studies deserve our attention. Schuckit et al. [3] studied half-siblings of alcoholics. The half-sibs share one biological parent with the known alcoholic, and there may be alcoholics among either their biological or their surrogate parents. Interviews disclosed that this was a sample enriched for alcoholics; there were 32 individuals diagnosed as alcoholic among the 164 half-siblings. Alcoholism was defined as "drinking in a manner that interferes with one's life." This study showed that it is the biological parent, not the foster parent, who affects the risk of alcoholism in the children. Of the alcoholic subjects, 62% had an alcoholic biological parent, whereas only 20% of the nonalcoholics had an alcoholic biological parent. Furthermore, a most interesting finding was that living with an alcoholic parent did not seem to increase the risk. Twenty-eight percent of the alcoholics and 22% of the nonalcoholics had lived for six years or more with an alcoholic parent figure (biological or surrogate).

Another recent family study is that of Goodwin and co-workers [4], who used Danish records to find a sample of 55 males who had been adopted by nonrelatives within six weeks of birth, and whose biological parent (usually the father) had been hospitalized for alcoholism. Controls were 78 men who had also been adopted in infancy and were matched for age but had no recorded alcoholic parents. Neither the interviewers nor data analysts knew of the subjects' parentage. It was shown that the foster homes into which these two groups were taken did not differ with respect to educational or economic level, or with respect to psychopathy. The subjects themselves did not differ between groups with respect to incidence of anxiety and depression or any kind of nonalcoholic psychopathy. (There was a higher incidence of divorce among the sons of alcoholics, but the divorces did not seem to be related to heavy drinking.) But the sons of alcoholics did have higher incidence of treatment for psychiatric conditions and more frequent hospitalization with psychiatric diagnoses than sons of nonalcoholics. Alcoholism was shown to contribute strongly to this difference. An "alcoholic" level of drinking was diagnosed from interview data on the basis of heavy drinking accompanied by alcohol-related problems, including social, legal, and medical consequences of drinking. The sons of alcoholics had greater histories of hallucinations, loss of control, morning drinking, and treatment for drinking problems than did the sons of nonalcoholics (Table 10-1). The overall incidence of alcoholism, according to the specified criteria, was significantly higher in the sons-of-alcoholics group (18% vs. 5%). Note that the numbers were small. There were only 10 alcoholics among the 55 sons of alcoholics, but this was a significantly higher proportion than the 4 alcoholics among the 78 controls. This important study again indicates that vulnerability to alcoholism is controlled in part by genes.

Table 10-1. Adopted-out sons of alcoholic and nonalcoholic parents[1]

	Proband	Control
Hallucinations	6*	0
Tremor	24	22
Morning drinking	29*	11
Delirium tremens	6	1
Rum fits	2	0
Any treatment for drinking	9*	1
Hospitalized for drinking	7	0
Drinking pattern		
Moderate drinker	51	45
Heavy drinker, ever	22	36
Problem drinker, ever	9	14
Alcoholic, ever	18*	5

1. The data show incidence of each event or pattern as percent within each group.

*$P < 0.05$ vs. the control group.

Is there a sex-linked determinant? Because of the higher incidence of alcoholism in males than females, the suggestion has often been made that any hereditary component of alcoholism might be sex linked. If the relevant defect were located on the X chromosome, it would be more often expressed in males, who carry only one such chromosome. The common occurrence of alcoholism in both father and son argues against this hypothesis, since fathers pass no X chromosomes to their sons. Nevertheless, some observations on color blindness seemed to support this notion. Cruz-Coke and co-workers reported that hospitalized alcoholics were color-blind more often than the general population, indicating an association with a known sex-linked trait [5]. This has been challenged on theoretical grounds—the two conditions would arise together only if they were extremely closely linked, or if the same gene product determined both traits (pleiotropy). Also, some investigators find that the color blindness is reversible and can probably be ascribed to the confusion of the hospitalized alcoholic, who is unable or unwilling to read the color charts.

Meanwhile, Kaij and Dock have done another study on the question of sex-linked inheritance of alcoholism [6]. This study was done entirely with records; no one was interviewed or seen by the investigator. Using data from the Swedish Temperance Boards, they identified individuals who had been registered there two generations ago. Tracing these families down to the present, Kaij and Dock found 270 of their grandsons over the age of 15. Recent Temperance Board records were used to find evidence for alcohol-related trouble among the grandsons. With these records, one could assess inheritance through the male and female lines separately, comparing the sons of sons with the sons of daughters of problem drinkers. If the condition were sex linked, it should pass through the female line and the sons of the daughters of alcoholics should have the higher rate of alcoholism. The results were that 21% of sons of sons were registered with Temperance Boards (28 out of 136 such subjects), and 18% of the sons of daughters were registered (24 out of 134). Thus it is clear that the inherited risk of alcoholism is not sex linked. Incidentally, the magnitude of the risk was here demonstrated again. Kaij and Dock calculated the percent registered

Fig. 10-2. *Alcohol problems run in families.* Incidental to a study of grandsons of alcoholics, the cumulated risk of being registered at the Swedish Temperance Board was calculated for 270 grandsons of men with alcohol problems (O—O) and for men in the general population (×—×). By 55 years of age, the frequency (F) of registration among the grandsons was three times that of the general population. From Kaij and Dock [6].

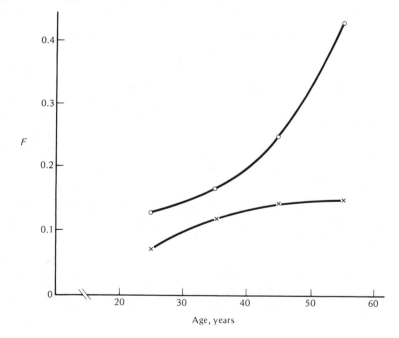

for each age group and found it much higher in the grandsons of alcoholics than in the general population, especially as they grew older (Fig. 10-2). As many as 43% of the grandsons of alcoholics had been registered by the age of 55, as against 15% in the general population. Again we find strong evidence that alcoholism runs in families, but this estimate of risk does not specifically implicate genetic influences.

Alcohol elimination rates

Sometimes it seems that the product of relevance and credibility is a constant. We have been considering data of the greatest clinical significance, but the findings depend on the diagnosis of alcoholism, sometimes with doubtful criteria, such as interviews with admittedly unreliable subjects. Now we turn to investigations where we can believe the numerical results but are less sure of the meaning. We begin with alcohol elimination rates. A small twin study by Vesell et al. [7] showed that monozygotic twin pairs had much less intrapair variance in their alcohol elimination rates than did dizygotic twin pairs. There were only seven pairs of each type, but the heritability was reported to be very high indeed (0.98). In a more recent study, Kopun and Propping [8] used a larger sample (40 twin pairs) and maintained more uniform conditions with respect to sex of subjects (all male), food and drug intake, etc. Their study also showed significant heritability, but of a lower degree (0.46).

Alcohol elimination rates have been examined in populations of different ethnic backgrounds. In particular, one would like to explain the high rates of alcoholism among American Indians and the reported low rates among Orientals. Two such studies have been conducted in Canada and one in the United States. They do not agree (Table 10-2). Fenna et al. [9] found that Indians and Eskimos metabolized ethanol more slowly than whites (suggesting that these native American populations might get drunker with a given dose). Within each of the three ethnic groups, the rate of elimination increased with the extent of heavy drinking in the subject's history, but the differences were small and were significant only for the American Indian group. A study by Reed et al. [10] had the opposite result. Whites metabolized the ethanol more slowly than the Indians, and in a third study [11], no differences were found.

A new interest in the genetics of alcohol effects has arisen since the publication of a paper by Wolff in 1972 [12], which describes strong dysphoric reactions to alcohol in Oriental subjects. Many Chinese and other

Table 10-2. Reported alcohol elimination rates in different populations[1]

| | Ethanol elimination rate, mg/ml per hour | | | |
	Whites	Indians	Eskimos	Chinese
Fenna	0.22 ± 0.018 (17)	0.16 ± 0.003 (26)	0.16 ± 0.001 (21)	–
Reed	0.12 ± 0.010 (37)	0.26 ± 0.036 (12)	–	0.13 ± 0.018 (15)
Bennion	0.16 ± 0.006 (30)	0.18 ± 0.007 (30)	–	–

1. The data are means and SEM for the number of subjects shown in parentheses. In Reed et al. [10], the Indians were Ojibwas and the Chinese were of Cantonese ancestry, living in Toronto. Indians in the Bennion and Li study [11] were "full blood" Indians from Arizona. The subjects in Fenna et al. [9] were from Edmonton.

Orientals report that they turn red and feel sick after drinking, even with small doses. Wolff studied the flush quantitatively, using a densitometer to measure light transmission through the earlobe at a wavelength for absorption by hemoglobin. The subjects were 78 Mongoloid people (Japanese, Taiwanese, and Koreans) and 34 Caucasians. Ethanol was given as beer. The Orientals were given about one beer and Caucasians two. In spite of the difference in dose, 83% of the Mongoloid subjects flushed, but only 6% of the Occidentals did. There were other reported symptoms, some of them quite unpleasant, as Table 10-3 shows. Four (18%) of the Oriental subjects fell asleep after only one beer. Wolff tested several newborn infants and found the same incidence of flushing, thereby demonstrating that diet and drinking history did not account for the difference between the populations. In addition, patients were questioned about reactions to alco-

Table 10-3. Symptoms of alcohol intoxication in Caucasoid and Mongoloid subjects[1]

Symptom	Caucasoid	Mongoloid
Hot in stomach	6	52
Palpitations	0	26
Tachycardia	3	44
Muscle weakness	3	26
Dizziness	9	37
Sleepiness	6	33
Falls asleep	0	18

1. The data are given as percent of the 34 white and 78 Oriental subjects who showed each symptom after a relatively small dose of ethanol.

hol among their parents. Many reported that their parents reacted as they did. Later, Wolff showed that American Indians also flush after drinking [13]. This would be expected, in view of the known origin of Indians from Mongoloid ancestors, but it makes interpretation of alcoholism rates very difficult. The same sort of unpleasant reaction that might account for low rates of alcoholism among the Chinese is also seen in a group with an extremely high prevalence of alcoholism, the North American Indians.

The flushed face suggests that acetaldehyde may be to blame for this unpleasant syndrome, since administration of acetaldehyde or its endogenous production in the alcohol-disulfiram reaction causes flushing. One sensible guess as to the mechanism is based on the observation that Orientals are much more likely than Occidentals to have the variant of alcohol dehydrogenase that is highly active at physiological pH (see Chapter 1). This enzyme form, called "atypical" in Switzerland where it was discovered was found in 85% of postmortem liver samples in Japan [14]. This is just the proportion of Oriental subjects who had the flushing reaction in Wolff's study. It was suggested that there is a period immediately after ingestion of ethanol when liver alcohol dehydrogenase, rather than NAD regeneration, is rate-limiting, since there may temporarily be plenty of NAD or oxidized substrates for NAD regeneration. Thus, individuals with an unusually active alcohol dehydrogenase might experience a brief rush of acetaldehyde soon after drinking. Recently, however, evidence is appearing that people who flush may differ from nonflushers with respect to metabolism of acetaldehyde, rather than ethanol. In an ingenious study [15], it was shown that one can obtain enough tissue from the roots of about 40 hairs plucked from an individual's head to separate isozymes of both alcohol dehydrogenase and aldehyde dehydrogenase by isoelectric focusing. Thus it was possible to correlate the enzyme patterns with history of flushing after drinking. Flushers lacked one of the isozymes of aldehyde dehydrogenase, thought to be the rate-limiting low-K_m enzyme. The presence or absence of atypical alcohol dehydrogenase was unrelated to flushing.

Animal studies

Genetic studies with human subjects are descriptive, not experimental. We observe the situation and ascribe some facets of it to genes. In order to do a genetic experiment, which involves manipulating genes and observing the result, we must go to animals. For this, mice are excellent subjects. They are fast-breeding little mammals that are close enough to humans for many purposes. Inbred, selectively bred, and heterogeneous stocks of mice have all been used in alcohol research [16]. Inbred strains are useful when one wants a stock of mice with essentially no genetic difference among individuals. Traits that differ among inbred strains are likely to be inherited. However, correlations of two or more traits among inbred strains are usually meaningless. If, for example, two strains differ in sensitivity to ethanol and also in their brain histamine levels, we cannot ascribe any more importance to this observation than to the apparently fortuitous association between alcohol sensitivity and coat color or tail length. If the correlation holds up over many inbred strains, it is much more credible. A reproducible stock of genetically heterogeneous mice, a large gene pool from which we can repeatedly sample for a long time, is extremely useful for studies of correlations of traits.

Selective breeding. Selective breeding, that is, outbreeding for some specific characteristic, is a powerful tool, especially when care is taken to retain the original diversity with regard to other characteristics. It is then legitimate to assume that traits accompanying the one that was selected for may be causally related to it. McClearn and Kakihana have developed two lines of mice that differ in "sleep time" after administration of hypnotic doses of ethanol [16]. The lines were derived from a large population of a well-defined, genetically heterogeneous stock, thus assuring a diversity of alleles. The duration of loss of righting reflex was measured after a standard dose of ethanol. Those males and females with the longest duration of loss of righting reflex ("longest sleepers") were mated, as were the male and female shortest sleepers. Thereafter, in each generation the longest-sleeping offspring of the long-sleep line were used for breeding, as were the shortest-sleeping offspring of the short-sleep line. This process was continued over many generations, with care taken to avoid inbreeding; thus, the genes that were selected were only those concerned with sleep time. The divergence of sleep times over 18 generations is shown in Figure 10-3. The mere fact that it is possible to breed for sleep time is evidence of genetic control of ethanol sensitivity. The slow divergence of the lines, rather than a quick Mendelian segregation into two or three populations, indicates that many genes are involved. The sleep times reflect differential brain sensitivity for ethanol, rather than metabolism, since ethanol elimination rates do not differ between these lines. The ED_{50} for loss of righting reflex was (at the 14th generation) 3.6 g/kg for the short-sleep and 1.6 g/kg for the more sensitive long-sleep line [17].

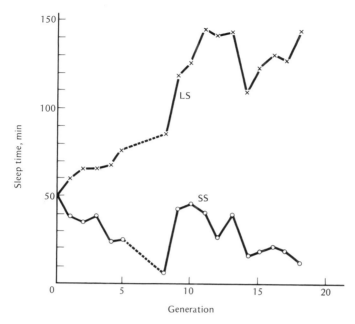

Fig. 10-3. *Divergence of long-sleep* (LS) *and short-sleep* (SS) *lines of mice.* Mice of a genetically heterogeneous stock were selectively bred by mating the most alcohol-sensitive individuals within the long-sleep line (×—×) and the most alcohol-resistant mice in the short-sleep line (O—O). (All available mice were bred in generations 6 and 7.) Alcohol sensitivity was determined by duration of loss of righting reflex after a standard dose of ethanol. From McClearn [16].

This gene-controlled difference in sensitivity of the brain to the hypnotic effect of ethanol does not necessarily apply to other actions of ethanol. For example, the LD_{50} for ethanol does not differ between the two lines, suggesting that death occurs by a different mechanism than hypnosis [18]. In addition, the lines do not differ in the effect of ethanol on rotarod performance [19]. Nevertheless, the genetic difference is not entirely specific for the loss of righting reflex, since these lines differ in their adrenal response to a stressful injection of ethanol [20]. Long-sleep mice responded to ethanol injections with a greater rise in plasma corticosterone than did short-sleep mice, but other stressful procedures such as histamine injection or electric shock did not cause a different reaction in the two strains. Figure 10-4 shows the ethanol-specific line difference.

Next, we can ask whether the lines differ in response to drugs other than

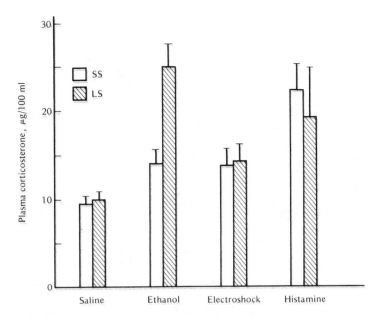

Fig. 10-4. *Adrenal response to stress in ethanol-sensitive and resistant lines of mice.*
Plasma corticosterone levels were measured 60 min after treatment with saline,
ethanol (1.6 g/kg i.p.), electroshock (0.4 mA footshock), or histamine (50 mg/kg
s.c.). Error bars represent the SEM for 8–12 mice. A difference between the LS
and SS mice is seen only with ethanol and not with other stressors, indicating
specificity of the genetic line difference. From Kakihana [20].

ethanol. Other alcohols (methanol, *n*-butanol, and *t*-butanol) affect the long-
sleep line more than the short-sleep, suggesting a common site of action.
But there was no line difference when other hypnotics such as pentobarbital or ether were given [18]. This observation is important because ethanol
and barbiturate actions, both acute and chronic, are so similar that one
might well suppose that they act at the same site. The lack of strain difference in barbiturate action tells us that barbiturates do not act through the
primary site of alcohol action. The secondary actions of both types of drug
(the result of perturbation at the primary site) must converge in a final
common path to produce the same behavioral syndrome.

A brief two-generation selective breeding experiment [21] sufficed to show
that alcohol withdrawal reactions in mice (scored by convulsions elicited by
handling) are strongly controlled by genes. Individual mice consistently
score high or low in this respect when tested repeatedly in successive cy-

cles of intoxication and withdrawal. This trait is clearly transmitted to their offspring.

Rats have also been selectively bred for differences in sensitivity to ethanol. The "least affected" and "most affected" lines were separated on the basis of the effect of alcohol on motor activity [22].

Rat strains have been selectively bred for differences in alcohol preference. It was already known that genes affect alcohol preference because there are large species differences [23]. When offered a choice of 10% alcohol or water as drinking fluid, hamsters will take 88% of their fluid as the ethanol solution, whereas guinea pigs take almost no alcohol. Both species can clearly differentiate between the two solutions. Rabbits, on the other hand, take about equal amounts of each, as if they could not perceive the taste or the effects of the ethanol. Eriksson's AA (drinker) and ANA (nondrinker) lines of rats [24] were selected by outbreeding for differences in ethanol consumption in a water/alcohol free-choice design. The alcohol-avoiding strain has higher levels of blood acetaldehyde after an injection of ethanol [25], which suggests that aversive effects of acetaldehyde may contribute to genetic differences in alcohol intake. The alcohol-preferring strain is less sensitive to acute alcohol intoxication, as measured by the tilt plane test for motor incoordination [26]. However, this difference in sensitivity is not drug specific. It applies not only to alcohols (ethanol, isopropanol, tertiary butanol), but also to barbital.

Inbred strains. The original observation that stimulated much work on mouse genetics in relation to alcohol was that inbred strains of mice differ substantially in their voluntary intake of alcohol [27]. C57BL mice take significantly more ethanol than other well-known strains such as DBA or BALB. Correlates of this preference have been explored. The drinker C57BL mice have a higher activity of liver alcohol dehydrogenase than nondrinker strains, but the *in vivo* elimination rates do not reflect this difference [28]. More important, apparently, is the activity of aldehyde-metabolizing enzymes. The C57BL mice have more aldehyde dehydrogenase activity than DBA, and this is reflected in lower blood concentrations of acetaldehyde during ethanol metabolism than in nondrinker strains [29]. The same is true of the drinker versus nondrinker rats studied in Finland, as noted above. It seems likely that acetaldehyde makes ethanol aversive to the nondrinker strain.

As in the AA/ANA strains of rats, the drinker mice are less sensitive to acute alcohol effects than nondrinkers. BALB mice have longer ethanol sleep times than C57BL, although alcohol elimination rates are the same in the two strains. C57BL mice regained their righting reflex at a mean brain level of 4.3 mg/g after a hypnotic dose of ethanol, while BALB mice continued to sleep until their brain ethanol concentration had fallen to 2.9 mg/g, a clear indication of relatively greater CNS sensitivity to ethanol in BALB mice [30]. The strain that was least affected in terms of sleep time was also least responsive to the stress effect of ethanol. Plasma corticosterone levels rose higher in DBA than in C57BL mice after a single low dose of ethanol [31].

The difference in alcohol sensitivity between inbred strains, like the previously discussed difference between outbred lines of mice, is specific for ethanol. C57BL mice are slightly more sensitive to pentobarbital than DBA mice (the opposite of the alcohol effect) [32]. It appears that ethanol and pentobarbital have different primary sites of action. The drugs initiate some change at their primary site and it is carried along a chain of neuronal events until the action reaches a common site, where both kinds of drug eventually produce the same type of behavior, e.g., ataxia. Genes can apparently act at multiple sites along the separate neuronal pathways.

Summary

Alcoholism runs strongly in families; well-designed studies of twins and adoptees indicate that some of this effect is inherited. It does not seem to be sex linked. Alcohol elimination rates vary among ethnic groups, but not consistently in different studies. Oriental people often flush after drinking; this may be due to slow metabolism of acetaldehyde. In animal studies, selective breeding for alcohol sensitivity has provided evidence that alcohols and barbiturates differ in their primary sites of action. Several aspects of an animal's response to ethanol (sensitivity, preference, adrenal activation) are susceptible to genetic manipulation in rats or mice.

References

1. Partanen, J., Bruun, K. and Markkanen, T. Inheritance of Drinking Behavior. Finnish Foundation for Alcohol Studies, Helsinki, 1966.
2. Kaij, L. Alcoholism in Twins. Studies on the Etiology and Sequels of Abuse of Alcohol. Almqvist and Wiksell, Stockholm, 1960.
3. Schuckit, M.A., Goodwin, D.A. and Winokur, G. A study of alcoholism in half siblings. Am. J. Psychiat. 128: 1132–1136, 1972.
4. Goodwin, D.W., Schulsinger, F., Hermansen, L., Guze, S.B. and Winokur,

G. Alcohol problems in adoptees raised apart from alcoholic biological parents. Arch. Gen. Psychiat. 28: 238–243, 1973.

5. Cruz-Coke, R. and Varela, A. Inheritance of alcoholism. Its association with colour-blindness. Lancet 2: 1282–1284, 1966.

6. Kaij, L. and Dock, J. Grandsons of alcoholics. A test of sex-linked transmission of alcohol abuse. Arch. Gen. Psychiat. 32: 1379–1381, 1975.

7. Vesell, E., Page, J.G. and Passananti, G.T. Genetic and environmental factors affecting ethanol metabolism in man. Clin. Pharmacol. Ther. 12: 192–201, 1971.

8. Kopun, M. and Propping, P. The kinetics of ethanol absorption and elimination in twins and supplementary repetitive experiments in singleton subjects. Eur. J. Clin. Pharmacol. 11: 337–344, 1977.

9. Fenna, D., Mix, L., Schaefer, O. and Gilbert, J.A.L. Ethanol metabolism in various racial groups. Can. Med. Assoc. J. 105: 472–475, 1971.

10. Reed, T.E., Kalant, H., Gibbins, R.J., Kapur, B.M. and Rankin, J.G. Alcohol and acetaldehyde metabolism in Caucasians, Chinese and Amerinds. Can. Med. Assoc. J. 115: 851–855, 1976.

11. Bennion, L.J. and Li, T.-K. Alcohol metabolism in American Indians and whites. Lack of racial differences in metabolic rate and liver alcohol dehydrogenase. New Eng. J. Med. 294: 9–13, 1976.

12. Wolff, P.H. Ethnic differences in alcohol sensitivity. Science 175: 449–450, 1972.

13. Wolff, P.H. Vasomotor sensitivity to alcohol in diverse Mongoloid populations. Am. J. Hum. Genet. 25: 193–199, 1973.

14. Stamatoyannopoulos, G., Chen, S.-H. and Kukio, M. Liver alcohol dehydrogenase in Japanese: high population frequency of atypical form and its possible role in alcohol sensitivity. Am. J. Hum. Genet. 27: 789–796, 1975.

15. Goedde, H.W., Agarwal, D.P. and Harada, S. Genetic studies on alcohol-metabolizing enzymes: detection of isozymes in human hair roots. Enzyme 25: 281–286, 1980.

16. McClearn, G.E. Genetics and the pharmacology of alcohol. Proc. VI Int. Cong. Pharmacol. 3: 59–66, 1975.

17. Heston, W.D.W., Erwin, V.G., Anderson, S.M., and Robbins, H. A comparison of the effects of alcohol on mice selectively bred for differences in ethanol sleep-time. Life Sci. 14: 365–370, 1974.

18. Erwin, V.G., Heston, W.D.W. and McClearn, G.E. Effect of hypnotics on mice genetically selected for sensitivity to ethanol. Pharmacol. Biochem. Behav. 4: 679–683, 1976.

19. Sanders, B. Sensitivity to low doses of ethanol and pentobarbital in mice selected for sensitivity to hypnotic doses of ethanol. J. Comp. Physiol. Psychol. 90: 394–398, 1976.

20. Kakihana, R. Adrenocortical function in mice selectively bred for different sensitivity to ethanol. Life Sci. 18: 1131–1138, 1976.

21. Goldstein, D.B. Inherited differences in intensity of alcohol withdrawal reactions in mice. Nature 245: 154–156, 1973.

22. Riley, E.P., Freed, E.X. and Lester, D. Selective breeding of rats for differences in reactivity to alcohol. An approach to an animal model of alcoholism. I. General procedures. J. Stud. Alc. 37: 1535–1547, 1976.

23. Arvola, A. and Forsander, O. Comparison between water and alcohol consumption in six animal species in free-choice experiments. Nature 191: 819–820, 1961.

24. Eriksson, K. Genetic selection for voluntary alcohol consumption in the albino rat. Science 159: 739–741, 1968.

25. Eriksson, C.J.P. Ethanol and acetaldehyde metabolism in rat strains genetically selected for their ethanol preference. Biochem. Pharmacol. 22: 2283–2292, 1973.

26. Malila, A. Intoxicating effects of three aliphatic alcohols and barbital on two rat strains genetically selected for their ethanol intake. Pharmacol. Biochem. Behav. 8: 197–201, 1978.

27. McClearn, G.E. and Rodgers, D.A. Differences in alcohol preference among inbred strains of mice. Quart. J. Stud. Alc. 20: 691–695, 1959.

28. Wilson, E.C. Ethanol metabolism in mice with different levels of hepatic alcohol dehydrogenase. In Biochemical Factors in Alcoholism, Maickel, R.P., ed. Pergamon Press, New York, 1967, pp. 115–124.

29. Sheppard, J.R., Albersheim, P. and McClearn, G. Aldehyde dehydrogenase and ethanol preference in mice. J. Biol. Chem. 245: 2876–2882, 1970.

30. Kakihana, R., Brown, D.R., McClearn, G.E. and Tabershaw, I.R. Brain sensitivity to alcohol in inbred mouse strains. Science 154: 1574–1575, 1966.

31. Kakihana, R., Noble, E.P. and Butte, J.C. Corticosterone response to ethanol in inbred strains of mice. Nature 218: 360–361, 1968.

32. Randall, C.L. and Lester, D. Differential effects of ethanol and pentobarbital on sleep time in C57BL and BALB mice. J. Pharmacol. Exp. Ther. 188: 27–33, 1974.

Reviews

Omenn, G.S. Alcoholism. A pharmacogenetic disorder. Mod. Probl. Pharmacopsychiat. 10: 12–22, 1975.

Belknap, J.K. Genetic factors in the effects of alcohol: neurosensitivity, functional tolerance and physical dependence. In Alcohol Tolerance and Dependence, Rigter, H. and Crabbe, J.C., Jr., eds. Elsevier/North-Holland, 1980, pp. 157–180.

11. Interaction of Ethanol with Neurotransmitter Systems

Neurochemical studies of alcohol's effects on the brain are easier to carry out than to interpret. Ethanol shares with many other drugs an ability to change almost every aspect of brain chemistry that we can examine. This should not surprise us. The first lesson of functional neurochemistry is the same as that of neuroanatomy: everything is connected to everything else. If ethanol had a specific effect on one of the known transmitters (blocking its synthesis, for example), we would probably know it by now, by an evident depletion of that transmitter in every experiment. However, secondary changes would arise immediately in all the other transmitter systems. Until we measured the pertinent (depleted) one, we would be in the dark as to the real mechanism of action. It follows that a primary action of ethanol on some transmitter as yet unknown to us would reveal itself by a myriad of responses in all the known systems. These indirect effects would be susceptible to modulation by variations in experimental conditions, such as stress, timing, etc., causing disagreement between laboratories. An attractive possibility is that ethanol has a primary action in disordering cell membrane structure (Chapter 4) and that other changes follow, but this is too easy a way out. Facile explanations can be advanced for almost anything that happens in the brain as long as we have only vague ideas of the relation between membrane disorder and the function of membrane-bound proteins. The hypothesis needs testing in detail.

This chapter summarizes experimental findings, choosing a few from the huge literature in an attempt to present the most reproducible data. The value of these observations will only be apparent at a much later time.

Acetylcholine

Release of acetylcholine. The neuromuscular junction affords an opportunity to examine drug effects on acetylcholine release and physiological response in the absence of other transmitters. Acting presynaptically at this site, high concentrations of ethanol increase the rate of spontaneous release of acetylcholine. The frequency of miniature endplate potentials, representing single vesicles released from the presynaptic terminal, rises in the presence of 0.5 M ethanol or more [1]. In stimulated preparations, e.g., rat phrenic nerve-diaphragm, ethanol increases the amplitude of endplate potentials, indicating an increased number of synaptic vesicles released in response to stimulation. Tolerance develops to the presynaptic effect of ethanol in the rat phrenic nerve-diaphragm [2]. Nerve-muscle preparations from rats chronically treated with ethanol are resistant to the action of ethanol *in vitro*, showing a smaller increase in frequency of miniature endplate potentials than controls at each ethanol concentration.

In the brain, different effects are seen. Ethanol shares with many other CNS depressant drugs the ability to decrease the rate of release of acetylcholine in the brain, with consequent elevation of the amount of acetylcholine present in the tissue. When ethanol is applied to electrically stimulated brain slices *in vitro*, the release of acetylcholine into the medium declines [3]. The same effect is produced *in vivo*. Injections of ethanol decrease the amount of acetylcholine released into cortical cups [4]. Tolerance develops rapidly in this system over the course of half an hour, despite the continued presence of ethanol in the blood and in the perfusion fluid, as Figure 11-1 shows. Chronic administration of ethanol also produces tolerance to the inhibitory effect of ethanol on release of acetylcholine.

Postsynaptic effects. A postsynaptic effect of alcohols is revealed in the neuromuscular junction by increased amplitude and duration of miniature endplate potentials. This is because the drugs slow the rate of decay of the potentials [5]. The potency of alcohols is related to their lipid solubility, and it is believed that the effects of alcohols on the neuromuscular junction are caused by changes in the membrane lipids.

In the brain, alcohol affects binding of acetylcholine to its receptors. For example, high concentrations of alcohols block binding of tritiated quinuclidinyl benzilate to muscarinic receptors *in vitro* [6]. Ethanol has an appreciable effect at 0.5 M, and increasing chain length increases the potency

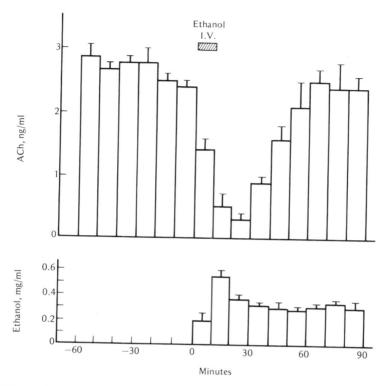

Fig. 11-1. *Ethanol blocks release of acetylcholine in the brain.* Acetylcholine (ACh) was measured in perfusion fluid collected over 10-minute periods in plastic cylinders sealed to the surface of the cerebral cortex in anesthetized cats. Bars show means and SEM for eight experiments. Ethanol (1 g/kg) was infused intravenously for 10 minutes, as shown. The lower histogram shows that ethanol appeared in the perfusate and its concentration remained stable while the response disappeared. From Sinclair and Lo [4].

of alcohols proportionally to the increase in their lipid solubility and membrane-disordering potency. Tolerance to ethanol in mice is accompanied by an increased number of muscarinic binding sites in some brain areas [7]. This effect is evident at the time of withdrawal, but disappears within a day.

Thus we get quite different messages from neuromuscular junction and from brain. In the former, a relatively simple system, ethanol seems to act directly on the membranes of synaptic vesicles or synapses, facilitating release of acetylcholine and the postsynaptic response to it. In the brain, acetylcholine release is decreased by ethanol, perhaps indirectly through some other neuronal system, and the postsynaptic membrane adapts by increasing its population of cholinergic receptors.

Serotonin

Acute ethanol. Many investigators have examined the brain concentrations of serotonin (5-hydroxytryptamine, 5-HT) after various doses of ethanol in rats, mice, and rabbits. The results are discrepant; most workers find no change in serotonin levels or turnover rates, but increases or decreases are seen by other investigators. Brain serotonin concentrations do not seem to be related to intoxication in any meaningful way.

Tolerance. Although there is likewise no clear message about brain serotonin metabolism after chronic administration of ethanol, there are interesting data implicating serotonin in the development of tolerance. When brain levels of serotonin were reduced to 5% of normal by prior administration of p-chlorophenylalanine (pCPA), the onset of tolerance was retarded, as measured by the performance of rats on the moving belt test for motor coordination [8]. Although pCPA did not modify the acute intoxication, it impaired performance of animals that had been chronically treated with ethanol. Depletion of serotonin also lessened tolerance to pentobarbital, and cross-tolerance between ethanol and pentobarbital, which was shown in these experiments with unusual clarity, was also retarded by pCPA. The primary effect of pCPA seemed to be an acceleration of the decay of tolerance, a process that presumably occurs continuously during development of tolerance but can best be measured after termination of ethanol treatment [9]. Figure 11-2 shows the effect of pCPA on the acquisition and loss of ethanol tolerance in rats. Other ways of manipulating brain serotonin concentrations, such as depleting with intracerebroventricular injections of 5,7-dihydroxytryptamine or augmenting with tryptophan treatments, had the expected effects. Tolerance to hypothermia (unlearned) was affected, as was tolerance to the impairment of learned behavior on the moving belt.

Norepinephrine

Norepinephrine turnover. As with serotonin, there is no general agreement about the effects of either acute or chronic ethanol administration on brain

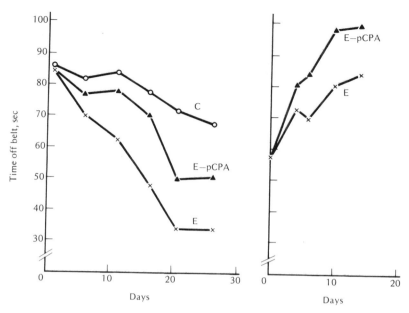

Fig. 11-2. *Depletion of brain serotonin (5-HT) blocks development of ethanol toler-ance, apparently by accelerating its decay.* The figure shows two separate experi-ments. Left, rats were treated daily with ethanol (E) and tolerance was demon-strated by decreased impairment over time. Impairment was measured as "time off belt" in the moving belt test described in Chapter 6. Less tolerance was seen in rats given pCPA along with the ethanol (E-pCPA). Controls (C), whether given pCPA or not, showed a small amount of tolerance due to the test dose of ethanol administered every few days. Right, recovery of rats previously made tolerant to ethanol. Separate groups of rats were tested at each time point to avoid any effect of the test doses of ethanol. Group E represents tolerant animals recovering with-out intervention. E-pCPA animals were given daily injections of pCPA during the time of recovery; they lost their tolerance more rapidly. Nontolerant rats (not shown) had impairment times of 90–95 seconds throughout. From Frankel et al. [8,9].

levels of norepinephrine. However, there is good evidence from several different kinds of experiments that alcohol administration increases the turnover of norepinephrine in brain. This is apparently not a direct effect of ethanol on adrenergic nerve terminals, since the opposite effect is seen *in vitro*. (High concentrations of ethanol reduce the rate of release of nor-epinephrine from electrically stimulated rat brain slices [3].) However, *in vivo*, ethanol treatment accelerates the decline in norepinephrine levels that occurs after administration of tyrosine hydroxylase inhibitors [10]; this indicates that ethanol activates noradrenergic neurons. Chronic administration of ethanol also produces an increase in norepinephrine turnover [11] and in the brain concentration of 3-methoxy-4-hydroxyphenylglycol, the main central metabolite of norepinephrine [12]. The shift from oxidized to reduced catecholamine products that occurs peripherally on administration of ethanol (Chapter 2) was not seen in brain. It seems likely that both acute administration of ethanol and withdrawal after chronic administration are stressful. In both situations, there is increased excretion of norepinephrine and epinephrine in the urine. The increased norepinephrine turnover in the brain seems likely to be a nonspecific response to stress, rather than a direct effect of ethanol.

Experimental manipulation of norepinephrine. Administration of α-methyltyrosine or adrenergic blockers to mice during the alcohol withdrawal reaction briefly intensifies the withdrawal convulsions [13]. Both α- and β-blockers have this effect. The transient effect of propranolol is shown in Figure 11-3. In the same system, reserpine has a severe and prolonged effect that is often lethal. Thus it appears that brain catecholamine pathways serve some sedative or anticonvulsant functions, as is known in other situations as well.

Much interest has been stimulated in the possible therapeutic use of propranolol to alleviate either acute intoxication or the withdrawal syndrome. This β-blocker might be expected to be useful because of the autonomic signs that accompany alcohol withdrawal, despite the fact that it intensifies withdrawal convulsions in mice (Fig. 11-3). Experimental evidence does not support a useful role of propranolol as an amethystic agent or alcohol antagonist. (The ancient Greeks attributed to amethyst gemstones the power to prevent intoxication.) In human subjects, propranolol aggravates several of the acute effects of ethanol [14]. This action is not necessarily ascribable to β-blockade, since both isomers of d,l-propranolol are local anesthetics with some sedative effect. Further, propranolol has yet another action: it inhibits alcohol dehydrogenase [15] and may thus potentiate effects of ethanol. This same action would have a "tapering-off" effect at the end of an alcoholic drinking binge, which might be the mechanism for the reported beneficial effect of racemic propranolol on tremor, heart rate, and blood pressure in patients undergoing alcohol withdrawal reactions [16]. These reports cannot always be confirmed, however, and the evaluation of propranolol for this purpose is incomplete. The dextro isomer, inactive as a β-blocker, does suppress alcohol withdrawal signs in

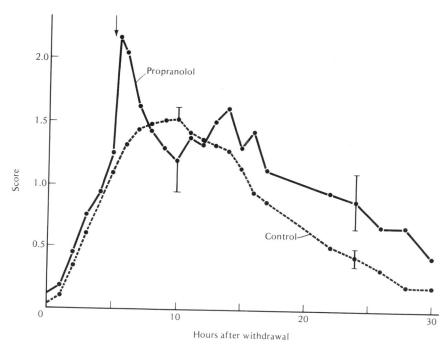

Fig. 11-3. *Propranolol intensifies alcohol withdrawal convulsions in mice.* Mice were made physically dependent on ethanol by three days of alcohol inhalation, using pyrazole to stabilize the blood alcohol concentration. On withdrawal, they were scored repeatedly for convulsions elicited by handling. The scores of mice that were untreated after withdrawal are shown by the broken line. The solid line represents mice injected with propranolol, 50 mg/kg i.p., at 5 hour after withdrawal (arrow). Bars show the standard error for 16 propranolol-treated mice and 95 controls. From Goldstein [13].

mice, presumably by a local anesthetic rather than an adrenergic mechanism [17].

Norepinephrine receptors. Chronic ethanol feeding reduces the number of β-adrenergic binding sites in rat or mouse brain and in rat heart, as shown by studies of binding of either tritiated dihydroalprenolol [18] or iodinated hydroxybenzylpindolol [7]. There is no change in the affinity of receptors for the ligands. Analysis of different kinds of β-receptor binding in brain indicates that the ethanol effect may be mostly at the sites designated as β-2 [7]. These are thought to be on nonneuronal cells, possibly blood vessels, and may respond to epinephrine under conditions of stress. Ethanol does not affect binding to adrenergic sites *in vitro* or after acute administration. Investigators disagree as to whether there is a rebound increase in binding capacity during the withdrawal reaction.

Dopamine

Brain levels of dopamine are generally found to be unchanged after acute alcohol administration, and there are conflicting reports about turnover of dopamine during alcohol intoxication, both acute and chronic.

Dopamine synthesis and release. Factors that control the synthesis and release of dopamine can be studied in rat striatal slices *in vitro*. In this system, depolarizing concentrations of potassium stimulate the synthesis of dopamine, apparently by an allosteric effect on tyrosine hydroxylase. After exposure to high potassium concentrations, the enzyme has an increased affinity for its pterin cofactor and decreased affinity for dopamine, its feedback inhibitor. This is not a direct effect of potassium, since the actual assay is carried out after the potassium has been washed out. Ethanol blocks this allosteric change; slices exposed to ethanol in high-potassium medium and subsequently assayed behave as though they had been in a low-potassium medium; i.e., the stimulatory effect of potassium on the tyrosine hydroxylase is not evident [19]. Tyrosine uptake into the slices and dopamine release from them were not affected by ethanol in this system, which is thought to be a model for the effects of *in vivo* activation of dopaminergic neurons. However, others have shown inhibition of dopamine release by ethanol *in vitro*, using somewhat different conditions.

Dopamine receptor sensitivity. Chronic ethanol treatment apparently does not change the binding of labeled spiroperidol or apomorphine [7,20]. However, there is some evidence for subsensitivity of dopaminergic systems; for example, chronic administration of ethanol blocks the stimulation of tyrosine hydroxylase that normally follows administration of neuroleptic drugs [21]. Striatal dopamine-sensitive adenylate cyclase is unchanged in sensitivity to dopamine at the time of withdrawal from ethanol [7], but may be subsensitive at 24 hours after ethanol withdrawal. Addition of ethanol to the *in vitro* assay restored the original sensitivity of the cyclase to dopamine [20]. Subsensitivity is also observed *in vivo*. The hypothermic response to the dopamine agonists apomorphine and piribedil is attenuated

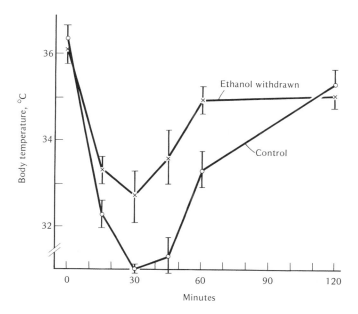

Fig. 11-4. *Decreased responsiveness to a dopamine agonist after chronic ethanol treatment.* The hypothermia that follows administration of apomorphine is shown for mice tested at 24 hours after withdrawal of ethanol (×) and in controls (○). The alcohol-treated animals were clearly less sensitive. Points are means and bars show SEM for seven to ten mice. From Rabin et al. [7].

after chronic treatment with ethanol (Fig. 11-4), but hypothermia after injection of clonidine, a norepinephrine agonist, is unaffected [7,21]. The brain contains dopamine receptors of different pharmacological specificity, both pre- and postsynaptic [22]. Their nomenclature is not yet standardized, and some of the discrepancies in the literature may result from failure to take this diversity into account.

Cyclic nucleotides

Adenosine 3':5'-cyclic monophosphate (cAMP). Some investigators find no change in cAMP concentration in any brain region, either with acute or chronic ethanol administration [23], but others see changes that vary among different parts of the brain [24]. It is unclear whether tolerance develops.

Ethanol has no direct effect on basal adenylate cyclase of brain *in vitro*, nor is there an effect of a single dose of ethanol on the basal or fluoride-stimulated activity of adenylate cyclase systems in brain tissue [25]. Nevertheless, adenylate cyclase systems in other tissues do sometimes respond to alcohol in ways that may reflect the drug's ability to disorder membrane bilayers (Chapter 4). For example, short chain aliphatic alcohols stimulate adenylate cyclase in adipocytes [26]. Propanol and butanol activate the basal enzyme activity and also the activity stimulated by sodium fluoride or a guanyl nucleotide, but ethanol and methanol, perhaps because they are weaker, only affect the stimulated enzyme. Studies of in vitro effects of alcohols on the basal cyclase activity generally agree that alcohols stimulate the enzyme slightly if they have any measurable effects at all. When an activator such as a guanyl nucleotide is added in vitro, an exaggerated stimulatory effect of alcohol is seen. There is no change in the affinity for ATP or guanyl nucleotide, but there seems to be an increased number of active sites, possibly due to exposure of more protein when the membrane has been perturbed by the alcohol. Many different substances activate adenylate cyclase in different tissues. Some may affect the catalytic unit directly, whereas others function by binding to other proteins in a coupled system. The membrane environment of such a compex will very likely affect its function, but it is premature to attempt a rational synthesis of the experimental data reported so far.

Experimental modification of plasma membrane lipids affects the adenylate cyclase activity. Sinensky et al. have studied a mutant Chinese hamster ovary (CHO) cell line that fails to regulate its cholesterol synthesis and therefore contains cholesterol at levels determined by the growth medium. Raising the membrane cholesterol increases the degree of order in the membranes, as shown by EPR techniques, and stimulates the adenylate cyclase [27]. The actual increased stiffness of the membrane, rather than a specific chemical effect of cholesterol, affects the enzyme. This effect is in the opposite direction from what one would expect on the basis of the direct disordering (and cyclase-stimulating) effects of alcohols in the brain and liver. In the cholesterol-enriched cultured CHO cells, the basal adenylate cyclase activity is elevated, but there is little further activation by prostaglandin or fluoride. This may be related to an observation of increased activity of adenylate cyclase in mouse brain and liver after chronic treatment with ethanol [28], a situation where membrane cholesterol may be increased (Chapter 4). Here the normal stimulatory effect of norepinephrine on the enzyme is reduced, again suggesting that the enzyme is already partially activated.

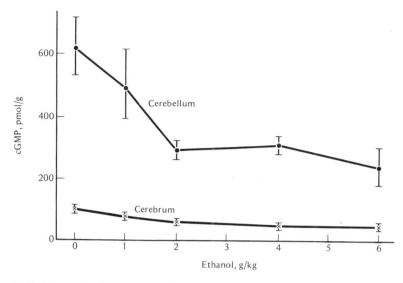

Fig. 11-5. *Dose-related decrease in brain cGMP after acute ethanol administration.* Rats were given ethanol orally and killed an hour later. The cGMP levels in various brain regions, two of which are shown here, were seen to fall in a dose-related way. Data are means and SEM for six rats per point. From Volicer and Hurter [23].

Guanosine 3':5'-cyclic monophosphate (cGMP). The brain concentrations of cGMP, unlike cAMP, consistently change after ethanol administration. Levels of cGMP are sharply decreased after single doses of ethanol, especially in the cerebellum. This effect is dose related, as Figure 11-5 shows, and reverses quickly as ethanol is eliminated. A similar decline in brain cGMP occurs after administration of other sedative drugs such as pentobarbital or diazepam. Partial tolerance to ethanol may develop in this system, and some workers observe a rebound in cerebellar cGMP after withdrawal [24], while others do not [29]. Basal guanylate cyclase activity is reported to be unaffected by acute, chronic, or *in vitro* administration of ethanol.

Gamma aminobutyric acid (GABA)

Brain levels. An early observation that brain GABA levels rise after acute administration of ethanol has proved difficult to reproduce, and other investigators have found no change or occasionally a decrease in brain GABA levels. Similarly, there is no agreement as to the effect of chronic ethanol treatment or withdrawal on brain levels of GABA. Reduced amounts of brain GABA are seen during or after chronic alcohol treatment, but it is not clear whether withdrawal of ethanol is followed by an increase or decrease in GABA. In one study [30], a 20% reduction in brain stem GABA seen after chronic ethanol treatment could be mimicked by administration of an inhibitor of glutamic acid decarboxylase (GABA synthesis). A 20% fall in GABA sufficed to produce audiogenic seizures in normal rats, suggesting that a deficiency of GABA contributes to the seizure susceptibility of the ethanol withdrawal period. Little is known about ethanol effects on the enzymes that synthesize and degrade GABA. There are only scattered reports, which generally do not agree.

Effect on withdrawal reactions. GABA and its agonists, injected intracerebroventricularly, will suppress audiogenic seizures during the alcohol withdrawal syndrome in rats [30]. The potency of agonists in suppressing the withdrawal seizures correlated with their ability to displace GABA from sodium-independent binding sites in brain tissue. Muscimol was by far the most potent in both systems. Drugs that raise brain GABA concentrations by blocking its breakdown also suppress alcohol withdrawal signs in mice and rats. Aminooxyacetic acid and ethanolamine-0-sulfate, inhibitors of GABA transaminase, have this effect. Sodium valproate, an anticonvulsant, is also effective [31]. It is an inhibitor of GABA transaminase and succinic semialdehyde dehydrogenase, but its action on GABA breakdown is now thought to be too weak to explain its anticonvulsant effect. Like ethanol, it can disorder membrane lipids. The mechanism of its anticonvulsant action has not been established.

Summary

This chapter summarizes reports of various neurochemical responses to ethanol. Amid the confusion and conflict in the literature, one can pick out only a few findings that have been replicated. Ethanol facilitates release of acetylcholine at the neuromuscular junction, but in the CNS it acts like other depressant drugs to block release of acetylcholine. The release of several other transmitters may be inhibited by ethanol, but some of this action may be offset by indirect effects of ethanol as a stressor. The increased norepinephrine turnover after acute ethanol administration or after withdrawal may be interpreted as stress effects. An interesting set of internally consistent data (not yet confirmed in other laboratories) indicates that serotonin plays an important and specific role in the development and decay of functional tolerance. Changes in sensitivity of brain receptors for their endogenous ligands are often sought and sometimes found. For example, some dopaminergic pathways seem to be subsensitive to dopamine after chronic ethanol treatment, and some β-adrenergic receptors may increase in abundance.

The actions of alcohol on the various components of the brain adenylate cyclase system are still in doubt, but the cGMP system may be simpler. Depressant drugs, including ethanol, reduce the brain levels of cGMP. It may prove possible to relate many of these changes to the membrane-disordering actions of ethanol. Since the brain functions by an almost unimaginably complex interplay of millions of synapses, it is not surprising that we cannot yet assign an action of a drug to a single neurotransmitter.

References

1. Gage, P.W. The effect of methyl, ethyl and n-propyl alcohol on neuromuscular transmission in the rat. J. Pharmacol. Exp. Ther. 150: 236–243, 1965.
2. Curran, M. and Seeman, P. Alcohol tolerance in a cholinergic nerve terminal: relation to the membrane expansion-fluidization theory of ethanol action. Science 197: 910–911, 1977.
3. Carmichael, F.J. and Israel, Y. Effects of ethanol on neurotransmitter release by rat brain cortical slices. J. Pharmacol. Exp. Ther. 193: 824–834, 1975.
4. Sinclair, J.G. and Lo, G.F. Acute tolerance to ethanol on the release of acetylcholine from the cat cerebral cortex. Can. J. Physiol. Pharmacol. 56: 668–670, 1978.
5. Gage, P.W., McBurney, R.N. and Schneider, G.T. Effects of some aliphatic alcohols on the conductance change caused by a quantum of acetylcholine at the toad endplate. J. Physiol. 244: 409–429, 1975.
6. Fairhurst, A.S. and Liston, P. Effects of alkanols and halothane on rat brain muscarinic and α-adrenergic receptors. Eur. J. Pharmacol. 58: 59–66, 1979.
7. Rabin, R.A., Wolfe, B.B., Dibner, M.D., Zahniser, N.R., Melchior, C. and Molinoff, P.B. Effects of ethanol administration and withdrawal on neurotransmitter receptor systems in C57 mice. J. Pharmacol. Exp. Ther. 213: 491–496, 1980.
8. Frankel, D., Khanna, J.M., LeBlanc, A.E. and Kalant, H. Effect of p-chlorophenylalanine on the acquisition of tolerance to ethanol and pentobarbital. Psychopharmacologia 44: 247–252, 1975.
9. Frankel, D., Khanna, J.M., Kalant, H. and LeBlanc, A.E. Effect of p-chlorophenylalanine on the loss and maintenance of tolerance to ethanol. Psychopharmacology 56: 139–143, 1978.
10. Hunt, W.A. and Majchrowicz, E. Alterations in the turnover of brain norepi-

nephrine and dopamine in alcohol-dependent rats. J. Neurochem. 23: 549–552, 1974.

11. Ahtee, L. and Svartström-Fraser, M. Effect of ethanol dependence and withdrawal on the catecholamines in rat brain and heart. Acta Pharmacol. Toxicol. 36: 289–298, 1975.

12. Karoum, F., Wyatt, R.J. and Majchrowicz, E. Brain concentrations of biogenic amine metabolites in acutely treated and ethanol-dependent rats. Br. J. Pharmacol. 56: 403–411, 1976.

13. Goldstein, D.B. Alcohol withdrawal reactions in mice: effects of drugs that modify neurotransmission. J. Pharmacol. Exp. Ther. 186: 1–9, 1973.

14. Alkana, R.L., Parker, E.S., Cohen, H.B., Birch, H. and Noble, E.P. Reversal of ethanol intoxication in humans: an assessment of the efficacy of propranolol. Psychopharmacology 51: 29–37, 1976.

15. Duncan, R.J.S. The inhibition of alcohol and aldehyde dehydrogenases by propranolol. Mol. Pharmacol. 9: 191–198, 1973.

16. Sellers, E.M., Zilm, D.H. and Degani, N.C. Comparative efficacy of propranolol and chlordiazepoxide in alcohol withdrawal. J. Stud. Alc. 38: 2096–2108, 1977.

17. Freund, G. Prevention of ethanol withdrawal seizures in mice by local anesthetics and dextro-propranolol. Adv. Exp. Med. Biol. 85A: 1–13, 1977.

18. Banerjee, S.P., Sharma, V.K. and Khanna, J.M. Alterations in β-adrenergic receptor binding during ethanol withdrawal. Nature 276: 407–409, 1978.

19. Bustos, G., Roth, R.H. and Morgenroth, V.H., III. Activation of tyrosine hydroxylase in rat striatal slices by K^+-depolarization—Effect of ethanol. Biochem. Pharmacol. 25: 2493–2497, 1976.

20. Tabakoff, B. and Hoffman, P.L. Development of functional dependence on ethanol in dopaminergic systems. J. Pharmacol. Exp. Ther. 208: 216–222, 1979.

21. Tabakoff, B., Hoffman, P.L. and Ritzmann, R.F. Dopamine receptor function after chronic ingestion of ethanol. Life Sci. 23: 643–648, 1978.

22. Seeman, P. Brain dopamine receptors. Pharmacol. Rev. 32: 230–313, 1980.

23. Redos, J.D., Hunt, W.A. and Catravas, G.N. Lack of alteration in regional brain adenosine-3′,5′-cyclic monophosphate levels after acute and chronic treatment with ethanol. Life Sci. 18: 989–992, 1976.

24. Volicer, L. and Hurter, B.P. Effects of acute and chronic ethanol administration and withdrawal on adenosine 3′:5′-monophosphate and guanosine 3′:5′-monophosphate levels in the rat brain. J. Pharmacol. Exp. Ther. 200: 298–305, 1977.

25. Kuriyama, K. and Israel, M.A. Effect of ethanol administration on cyclic 3′,5′-adenosine monophosphate metabolism in brain. Biochem. Pharmacol. 22: 2919–2922, 1973.

26. Stock, K. and Schmidt, M. Effects of short-chain alcohols on adenylate cyclase in plasma membranes of rat adipocytes. Naunyn-Schmiedeberg's Arch. Pharmacol. 302: 37–43, 1978.

27. Sinensky, M., Minneman, K.P. and Molinoff, P.B. Increased membrane acyl chain ordering activates adenylate cyclase. J. Biol. Chem. 254: 9135–9141, 1979.

28. Kuriyama, K. Ethanol-induced changes in activities of adenylate cyclase, guan-

ylate cyclase and cyclic adenosine 3',5'-monophosphate dependent protein kinase in the brain and liver. Drug Alc. Dep. 2: 335–348, 1977.

29. Hunt, W.A., Redos, J.D., Dalton, T.K. and Catravas, G.N. Alterations in brain cyclic guanosine 3':5'-monophosphate levels after acute and chronic treatment with ethanol. J. Pharmacol. Exp. Ther. 201: 103–109, 1977.

30. Cooper, B.R., Viik, K., Ferris, R.M. and White, H.L. Antagonism of the enhanced susceptibility to audiogenic seizures during alcohol withdrawal in the rat by γ-aminobutyric acid (GABA) and "GABA-mimetic" agents. J. Pharmacol. Exp. Ther. 209: 396–403, 1979.

31. Goldstein, D.B. Sodium bromide and sodium valproate: effective suppressants of ethanol withdrawal reactions in mice. J. Pharmacol. Exp. Ther. 208: 223–227, 1979.

Review

Tabakoff, B. and Hoffman, P.L. Alcohol and neurotransmitters. *In* Alcohol Tolerance and Dependence, Rigter, H. and Crabbe, J.C., Jr., eds. Elsevier/North-Holland, Amsterdam 1980. pp. 201–226.

12. Alcohol and the Endocrine System

Since ethanol affects a great variety of biochemical processes, we may expect it to cause multiple changes in the endocrine system. Ethanol effects have indeed been found in every hormone system where they have been sought. There are substantial bodies of data on ethanol as a stressor, on hypogonadism and feminization in males, and on vasopressin and oxytocin. This is an area where great advances have been made in methods (such as radioimmunoassays) and where research effort is overdue. As yet we lack any unifying theme for these endocrine effects, but they are of great clinical importance and pharmacological interest in themselves.

Adrenal cortical hormones

Stress effect of acute alcohol administration. Even before it was possible to measure levels of adrenal steroid hormones in plasma or urine, experimental data implicated the adrenal cortex in the actions of ethanol. Stressful conditions deplete the adrenal of ascorbic acid and cholesterol, and these compounds were assayed as a clue to the state of the gland. Single doses of ethanol reduced the amount of both compounds in the rat adrenal, and hypophysectomy abolished the effect, indicating that the ethanol acted at the pituitary level or above, rather than directly on the adrenal [1]. When assays for steroid hormones became available, it could be shown that single doses of ethanol increase the plasma level of cortisol in human subjects. Patients with pituitary lesions did not show this effect.

Animal studies reveal details of the actions of alcohol on the hypothalamic-pituitary-adrenal axis. Acute administration of ethanol causes a dose-related rise in plasma corticosteroid levels in rats [2]. Figure 12-1 shows the time course of this response in relation to the concentration of ethanol in the blood. Ethanol mimics adrenocorticotropic hormone (ACTH) in that an *in vivo* dose stimulates adrenal production of steroids; the increased rate of synthesis can be measured *in vitro* in excised glands [3]. In mice, alcohol injections deplete pituitary ACTH, and dexamethasone blocks the alcohol-induced rise in plasma steroid, presumably by shutting off ACTH release [4]. Ethanol does not change the rate of elimination of plasma glucocorticosteroids.

Genetic studies suggest that the pituitary-adrenal response to ethanol is closely related to alcohol's actions on the CNS. The magnitude of the adrenal response correlates with sensitivity to intoxicating effects of ethanol in genetically different populations of mice. Among inbred mouse strains, the ethanol-sensitive DBA mice show a stronger adrenal response than do C57BL mice that are relatively resistant to CNS effects of ethanol [5]. Ka-

Fig. 12-1. *Rise in plasma corticosterone after a single dose of ethanol.* Rats were treated with ethanol, 2 g/kg i.p. Plasma corticosterone ($\times$—$\times$) was measured at intervals in six to ten rats. Blood ethanol ($\bullet$—$\bullet$) values are for three to five animals. Means and SEM are shown. From Ellis [2].

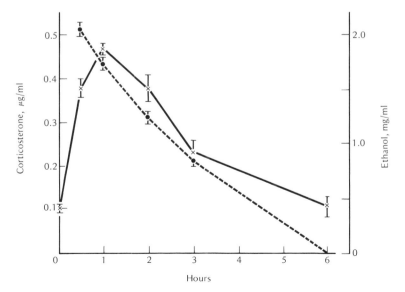

kihana [6] has tested the long-sleep (LS) and short-sleep (SS) lines of mice that differ in sensitivity to hypnotic effects of ethanol (see Chapter 10). The alcohol-sensitive (LS) mice had the greater plasma corticosterone rise after an ethanol injection. The line difference was specific for ethanol; other stressors such as electroshock, histamine, or saline injection did not affect one line more than the other. This is as it should be, since the selective breeding should have separated the lines only on the basis of their response to ethanol.

Chronic ethanol. Some workers report that tolerance does not develop to the pituitary-adrenal effect of ethanol. In human alcoholics, plasma cortisol may remain high over long periods of drinking; however, the condition of these patients is often complicated by stressful events such as trauma. Furthermore, low rates of cortisol elimination have been observed in alcoholic subjects, perhaps because of poor liver function. This might mask the development of tolerance at the adrenal level. In an experimental study with four alcoholic subjects [7], the serum cortisol remained high during 11–29 days of alcohol administration, with no evidence of adrenal exhaustion. Serum cortisol levels rose and fell roughly in parallel with blood alcohol concentrations and with the degree of intoxication.

Some animal studies agree that no tolerance develops to the adrenal steroid response. In Ellis's experiments [2] and those of Tabakoff and co-workers [8], plasma levels of corticosterone continued to be elevated after a week of ethanol administration to rats or mice, respectively, although there was a slight (nonsignificant) reduction in the response to ethanol. However, experiments of longer duration clearly showed the development of tolerance to the pituitary-adrenal effect of ethanol [4]. Mice that had been given alcohol in their drinking water for a few weeks were challenged with an injection of ethanol. A rise in plasma corticosterone was seen at first, but was progressively attenuated over six weeks of alcohol administration and stayed low for two to four weeks after withdrawal. Neither steroid nor ethanol metabolism had changed during the chronic treatment with ethanol, nor had the ability of the adrenals to respond to ACTH.

In DBA mice, chronic ethanol intake flattens the normal circadian rhythm of steroid release [9]. Neither the normal morning dip nor the afternoon peak in plasma corticosterone was seen in alcohol-treated mice; they had plasma corticosterone levels above controls at 10:30 hr and below at 18:30 hr (Fig. 12-2). This important finding suggests an explanation for many discrepant results in alcohol research. Observations made at different times

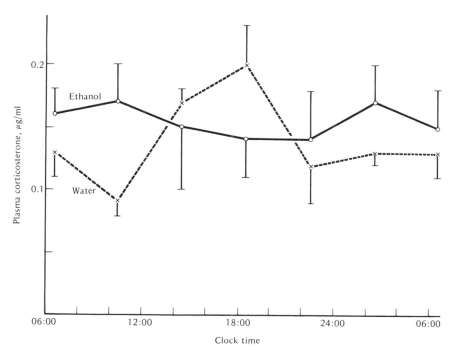

Fig. 12-2. *Flattening of circadian rhythm of corticosterone levels after chronic ethanol administration.* Mice were given ethanol in their drinking water for 15 weeks and were then killed at different times of day for plasma steroid assays. The points show means and SEM for seven to ten mice. Lights were on in the mouse room from 06:45 to 18:45. Mean blood alcohol levels in the alcohol group were 1.0 mg/ml at 06:30 and 0.3–0.5 mg/ml at the other times. From Kakihana and Moore [9].

of day might show opposite effects of a drug if the agent had decreased the amplitude or shifted the phase of a physiological circadian rhythm.

Adrenal steroids as modulators of actions of ethanol. Exogenously administered steroids can increase the rate of ethanol elimination. Repeated doses of cortisol over a six-day period increased the rate almost two-fold in rabbits [10]. A single dose had no effect. In dogs, administration of glucocorticoids such as dexamethasone and methyl prednisolone increased the rate of alcohol elimination by about 30% [11]. A considerable portion of the observed metabolic tolerance to ethanol may thus be mediated by glucocorticoids.

There are also reports in the literature, as yet unconfirmed, that adrenal steroids modulate some aspects of alcohol tolerance and physical dependence. Such experiments are difficult to control because dependence (and probably tolerance) is a dose related phenomenon. Its magnitude varies with the total exposure to alcohol (the product of blood ethanol concentration and the duration of alcohol intake). It is difficult to maintain two groups of animals at the same blood alcohol level when they metabolize the drug at different rates, which may happen if their endocrine systems are manipulated. Without monitoring of blood alcohol concentrations during chronic ethanol administration, we cannot be sure whether alcohol directly affects the processes of tolerance and physical dependence.

Gonadal hormones

It is a common clinical observation that male alcoholics with cirrhotic livers show signs of hypogonadism and feminization, such as testicular atrophy, decreased size of prostate and seminal vesicles, decreased facial and body hair, and gynecomastia. Their plasma testosterone levels are generally low. The mechanisms for these changes are currently being worked out in a series of exciting clinical and animal studies. The recent availability of radioimmunoassays has enormously accelerated progress in this area. It is now clear that ethanol affects gonadal hormones by a variety of different mechanisms. The story now unfolding is reminiscent of earlier attempts to find the mechanism of the acute fatty liver, where it turned out that almost every possible reaction in lipid metabolism was affected by ethanol.

Testosterone. Single doses or continued administration of ethanol reduce the levels of plasma testosterone both in male rats and in men. The current controversies about the site of action of ethanol in lowering plasma testosterone levels will probably be resolved by concluding that several mechanisms act simultaneously, with predominance of one or another depending on experimental conditions.

Ethanol may block the testicular synthesis of testosterone. A possible mechanism is suggested by the observation that alcohol dehydrogenase exists in rat testis, where it may mediate a decreased ratio of NAD to NADH in the presence of ethanol [12]. It remains to be demonstrated whether this effect is of sufficient magnitude to slow down the NAD-requiring steps in steroid biosynthesis. Formation of retinal by testicular alcohol dehydrogenase is a necessary step in spermatogenesis; the observed low sperm counts and germ cell injury in alcoholics might be caused by competition of ethanol and retinol for the small amount of available alcohol dehydrogenase [13]. However, effects deriving from alcohol dehydrogenase should be readily reversed as soon as the ethanol has been metabolized.

Other evidence that ethanol acts directly on testicular steroidogenesis is provided by studies of isolated, perfused rat testis [14]. In this preparation, added ethanol reduces gonadotropin-stimulated testosterone synthesis. Acetaldehyde has the same effect. Both compounds were used at high concentrations (Fig. 12-3). In intact rats the response of plasma testosterone to gonadotropin is also blocked by single doses of ethanol [15]. A relatively small dose (1 g/kg) halved the response to gonadotropin. The ethanol effect appeared noncompetitive, since it could not be overcome with maximal

Fig. 12-3. *Ethanol blocks gonadotropin-stimulated testosterone production in the testis.* The total testosterone of the isolated, perfused rat testis was measured as the sum of the hormone in the gland and that released into the perfusate. The bars at the left show unstimulated testes; there was no effect of ethanol under these conditions. Human chorionic gonadotropin (hCG) greatly stimulated testosterone production, and this effect was blocked by ethanol (3 mg/ml) or acetaldehyde (14 μg/ml). From Cobb et al. [14].

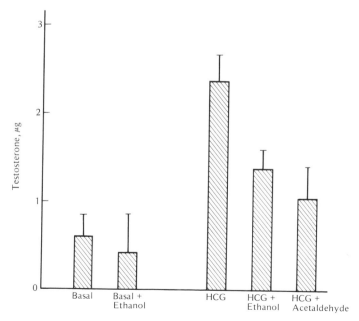

doses of chorionic gonadotropin. In men, the functional state of the testis can be examined by administering human chorionic gonadotropin to stimulate testosterone production. This response is greatly reduced in some alcoholic men [16].

Changes in testicular function are by no means the whole story. The liver also contributes to the reduction in circulating testosterone. The induction of microsomal enzymes by ethanol, discussed in Chapter 1, affects metabolism of steroid hormones. Hepatic 5-α-A-ring reductase has been shown to increase in rats after chronic alcohol administration; this would accelerate testosterone catabolism. In human volunteers, reductase activity was more than doubled after four weeks of experimental alcohol consumption [17]. However, alcohol intake over a period of years, by alcoholic patients or experimentally treated baboons, produces a decrease in activity of this hepatic enzyme [18]. The reductase converts testosterone to dihydrotestosterone, itself an active androgen in some tissues. The net effect of changes in its activity is not yet clear.

In the face of these various kinds of reduction in circulating free testosterone, the levels of plasma luteinizing hormone (LH) should tell us whether the pituitary and hypothalamus are functioning properly. A normal pituitary would respond to low plasma testosterone by putting out extra LH. Studies of LH levels when plasma testosterone has been reduced by ethanol have given contradictory results. Plasma LH sometimes rises after single doses of alcohol, indicating a hypothalamic-pituitary response to the fall in circulating testosterone. However, other investigators report that LH levels are inappropriately low or normal despite the urgent signal provided by a low testosterone. For example, in an experiment where chronic administration of ethanol to normal subjects produced a progressive fall in plasma testosterone, LH levels were variable, rather than responding strongly to the low testosterone [17]. Because ethanol did not prevent the pituitary response to luteinizing hormone-releasing hormone (LHRH), it probably acted at the hypothalmic level, blocking release of LHRH [19].

In rats, both testosterone and LH have been shown to fall after a large single dose of ethanol (Fig. 12-4). A biphasic effect of ethanol was described; there was a rise in both plasma testosterone and LH after low doses of alcohol and a fall in both hormones at moderate or high doses [20]. Ethanol blocks the rise in plasma LH that normally follows castration in the rat [15]; this clearly demonstrates the failure of the pituitary to respond to a scarcity of testosterone. The ethanol effect could be reversed by administration of LHRH, which puts a site of action in the hypothalamus,

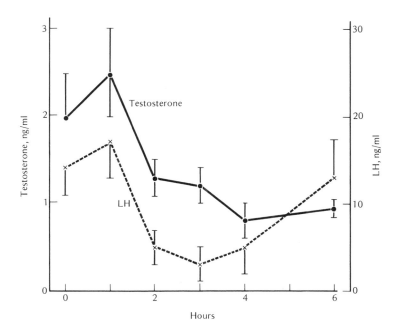

Fig. 12-4. *Fall in plasma testosterone and luteinizing hormone (LH) in rats after a single dose of ethanol.* Points and bars indicate the mean and SEM for groups of eight rats given ethanol, 2.5 g/kg i.p. and sampled at various times thereafter. Control data (shown at zero time) did not change over 6 hours after an injection of distilled water. From Cicero and Badger [20].

since (in this respect at least) the pituitary itself seems to be functioning properly.

Estrogens. The evident feminization of male alcoholics with liver disease was earlier assumed to be caused by failure of the liver to metabolize estrogens. However, as with testosterone, the situation now appears much more complicated. Some investigators have found an increase in plasma estradiol in such patients, but most have not. These male patients do have increased levels of estrone, a weaker estrogen than estradiol [21].

The investigation of effects of ethanol on reproductive hormones has concentrated on males so far, but it is entirely likely that equally damaging and equally complex effects on the sex hormones of women will be found when they are looked for. In laboratory animals, studies of the fetal alcohol syndrome frequently show that dams treated with alcohol are unable to conceive or to carry a pregnancy to term. The hormonal basis for this is unknown. In female rats after several weeks of chronic alcohol administration, the ovaries were reduced to less than half normal mass and lacked well-developed follicles and corpora lutea [22]. Estrogen target organs such as the uterus and vagina showed histological evidence of hormone deprivation. Plasma hormone levels were deranged in various ways (low levels of estradiol and progesterone, and high levels of two hormones that can be made in the adrenal, estrone and corticosterone).

Peptide hormones

Vasopressin. The evident diuretic effect of alcoholic beverages is caused by ethanol itself, not just by ingestion of large volumes of fluid [23]. That the effect is central was shown by Van Dyke and Ames [24], who produced diuresis in dogs by injecting ethanol into the carotid artery, in amounts far too small to have an effect by any other route (see Fig. 12-5). In the kidney itself, ethanol does not affect glomerular filtration or block the renal action of vasopressin. Its positive effect on free water clearance is caused by its ability to inhibit release of vasopressin, a posterior pituitary peptide hormone necessary for the resorption of water by the kidney tubules. Thus, ethanol does not affect urine volume when vasopressin release has already been turned off by a water load or in situations where no hormone is available, as in diabetes insipidus. It prevents the antidiuresis caused by saline administration or by cholinergic agents (acetylcholine or nicotine) that act at the hypothalamic supraoptic nuclei where vasopressin is synthesized.

Tolerance develops very rapidly. Alcohol diuresis is transient even if high blood alcohol levels are maintained by prolonged or repeated alcohol administration [23,24].

Oxytocin. Oxytocin, a posterior pituitary peptide closely related to vasopressin, promotes uterine contraction during labor, as well as in response to suckling. Uterine motility can be measured in rabbits or in women, either externally or (more accurately) by measuring the pressure in a balloon inserted in the uterus. Ethanol suppresses uterine motility at term or in response to suckling, but it does not suppress contractions induced by administration of exogenous oxytocin [25]. Thus, the ethanol effect is on the pituitary or hypothalamus, not on the peripheral organ. Ethanol also suppresses spontaneous uterine contractions occurring in nonpregnant women, especially the strong, regular contractions during menstruation.

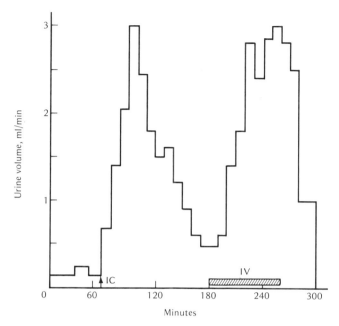

Fig. 12-5. *The diuretic action of ethanol is central.* The urine flow in an unanesthetized dog was measured before and after injecting a very small dose of alcohol into the carotid artery (IC) or infusing a much larger dose intravenously (IV). At 60 minutes, 0.05 g/kg was injected IC at 0.01 g/kg per minute. At 180–260 minutes, 2.07 g/kg was infused IV at 0.026 g/kg per minute. Similar effects on urine flow were observed. From Van Dyke and Ames [24].

Oxytocin participates in milk ejection response to suckling. The young may receive little or no milk if oxytocin release has been blocked by ethanol. For example, a litter of rabbits that received 38 g of milk during normal suckling obtained only 4 g when the mother had been treated with a moderate dose of ethanol [25]. This aspect of alcohol's effect on the newborn has received less attention than it deserves.

The ability of ethanol to postpone parturition in pregnant rabbits led to its therapeutic use for women in premature labor. A sequential clinical trial clearly showed that ethanol infusions were more effective than glucose infusions in postponing labor [26]. Ethanol is not effective after the membranes have ruptured or when the cervix is dilated. A substantial dose is needed—usually 1.2 g/kg given over a two-hour period, followed by frequent small doses during the next six to ten hours [27]. This is enough to cause obvious intoxication of the mother, often with nausea, vomiting, and the possibility of aspiration; these are serious disadvantages. If the treatment fails to prevent delivery, the infant is intoxicated at birth and may remain so for some time, since neonates have little capacity for ethanol metabolism. Long-term effects of such treatment have not been investigated.

Tolerance affected by peptides. Several groups of investigators report that peptide hormones such as vasopressin can improve memory. Behavioral effects are produced by fragments of several anterior or posterior pituitary hormones that are themselves devoid of hormone action. Facilitation of acquisition and delay of extinction of learned behavior may be caused by these compounds [28]. Facilitation of morphine tolerance and dependence has been reported, and similar results have recently been obtained with respect to ethanol tolerance. Hoffman and co-workers [29] treated mice with vasopressin or oxytocin during chronic administration of ethanol and after withdrawal. Tolerance developed in all groups, as shown by the degree of hypothermia and duration of loss of righting reflex after a challenge dose of alcohol, but the tolerance was maintained much longer in the vasopressin groups than in mice treated with oxytocin or in controls. Signs of physical dependence were not affected by either peptide.

Such results are no doubt just the beginning of a set of investigations on the relation of ethanol to the newly discovered peptides of the brain and pituitary, including the endorphins. Relationships between alcohol and opiate addiction have seemed doubtful to me because of the obvious differences in the patterns of intoxication and withdrawal reactions of the two drug types. Discovery of receptors for opiates reinforced the belief that the drugs had little in common biochemically, since chemical specificity seems unlikely in the actions of ethanol. However, the discovery of the endorphins, which may have a range of pharmacological effects quite different from those of the opiates, makes it more reasonable to accept some similarity of opiate and ethanol effects. If endorphins and other peptides function as mediators or modulators of synaptic transmission, we may expect to hear about a variety of interactions between peptides and ethanol, just as between catecholamines and ethanol. Some may be directly related to the mode of action of ethanol and others may be red herrings. Reversal of ethanol effects by naloxone has already been reported and also denied.

Summary

In its role as stressor, ethanol acts on the hypothalamic-pituitary-adrenal axis to raise plasma levels of glucocorticoids. It is not clear whether tolerance develops with chronic ethanol administration, but there is evidence that such treatment damps the normal circadian rhythm of adrenal steroid function. Corticosterone, in turn, accelerates the elimination of ethanol. Ethanol causes a decrease in plasma testosterone, for which many mechanisms are coming to light. Testicular synthesis of testosterone is inhibited, hepatic androgen metabolism is stimulated, and the hypothalamic-pituitary response to low levels of circulating testosterone is blunted. Further, estrogen metabolism in both males and females is deranged by ethanol. Release of posterior pituitary peptide hormones is blocked by ethanol. This causes the well-known diuretic effect as well as inhibition of uterine motility and milk ejection. The behavioral effects of pituitary peptides may include modification of ethanol tolerance.

References

1. Forbes, J.C. and Duncan, G.M. The effect of acute alcohol intoxication on the adrenal glands of rats and guinea pigs. Quart. J. Stud. Alc. 12: 355–359, 1951.
2. Ellis, F.W. Effect of ethanol on plasma corticosterone levels. J. Pharmacol. Exp. Ther. 153: 121–127, 1966.
3. Kalant, H., Hawkins, R.D. and Czaja, C. Effect of acute alcohol intoxication on steroid output of rat adrenals in vitro. Am. J. Physiol. 204: 849–855, 1963.
4. Noble, E.P. Ethanol and adrenocortical stimulation in inbred mouse strains. In Recent Advances in Studies of Alcoholism, Mello, N.K. and Mendelson, J.H., eds. NIMH/NIAAA, Rockville MD, 1971.
5. Kakihana, R., Noble, E.P. and Butte, J.C. Corticosterone response to ethanol in inbred strains of mice. Nature 218: 360–361, 1968.
6. Kakihana, R. Adrenocortical function in mice selectively bred for different sensitivity to ethanol. Life Sci. 18: 1131–1138, 1976.
7. Mendelson, J.H., Ogata, M. and Mello, N.K. Adrenal function and alcoholism. I. Serum cortisol. Psychosom. Med. 33: 145–157, 1971.
8. Tabakoff, B., Jaffe, R.C. and Ritzmann, R.F. Corticosterone concentrations in mice during ethanol drinking and withdrawal. J. Pharm. Pharmacol. 30: 371–374, 1978.
9. Kakihana, R. and Moore, J.A. Circadian rhythm of corticosterone in mice: the effect of chronic consumption of alcohol. Psychopharmacologia 46: 301–305, 1976.
10. Fischer, H.-D. Zum Einfluss von Hydrocortison auf die Entgiftungsgeschwindigkeit des Äthanols. Biochem. Pharmacol. 15: 785–791, 1966.

11. Clark, W.C. and Owens, P.A. The influence of glucocorticoid, epinephrine and glucagon on ethanol metabolism in the dog. Arch. Int. Pharmacodyn. 162: 355–363, 1966.

12. Van Thiel, D.H. and Lester, R. Alcoholism: its effect on hypothalamic pituitary gonadal function. Gastroenterology 71: 318–327, 1976.

13. Van Thiel, D.H., Gavaler, J. and Lester, R. Ethanol inhibition of vitamin A metabolism in the testes: possible mechanism for sterility in alcoholics. Science 186: 941–942, 1974.

14. Cobb, C.F., Ennis, M.F., Van Thiel, D.H., Gavaler, J.S. and Lester, R. Isolated testes perfusion: a method using a cell- and protein-free perfusate useful for the evaluation of potential drug and/or metabolic injury. Metabolism 29: 71–79, 1980.

15. Cicero, T.J., Meyer, E.R. and Bell, R.D. Effects of ethanol on the hypothalamic-pituitary-luteinizing hormone axis and testicular steroidogenesis. J. Pharmacol. Exp. Ther. 208: 210–215, 1979.

16. Van Thiel, D.H., Lester, R. and Sherins, R.J. Hypogonadism in alcoholic liver disease: evidence for a double defect. Gastroenterology 67: 1188–1199, 1974.

17. Gordon, G.G., Altman, K., Southren, A.L., Rubin, E. and Lieber, C.S. Effect of alcohol (ethanol) administration on sex-hormone metabolism in normal men. New Eng. J. Med. 295: 793–797, 1976.

18. Gordon, G.G., Vittek, J., Ho, R., Rosenthal, W.S., Southren, A.L. and Lieber, C.S. Effect of chronic alcohol use on hepatic testosterone 5-α-A-ring reductase in the baboon and in the human being. Gastroenterology 77: 110–114, 1979.

19. Leppaluoto, J., Rapeli, M., Varis, R. and Ranta, T. Secretion of anterior pituitary hormones in man: effects of ethyl alcohol. Acta Physiol. Scand. 95: 400–406, 1975.

20. Cicero, T.J. and Badger, T.M. Effects of alcohol on the hypothalamic-pituitary-gonadal axis in the male rat. J. Pharmacol. Exp. Ther. 201: 427–433, 1977.

21. Kley, H.K., Nieschlag, E., Wiegelmann, W., Solbach, H.G. and Krüskemper, H.L. Steroid hormones and their binding in plasma of male patients with fatty liver, chronic hepatitis and liver cirrhosis. Acta Endocrinol. 79: 275–285, 1975.

22. Van Thiel, D.H., Gavaler, J.S., Lester, R. and Sherins, R.J. Alcohol-induced ovarian failure in the rat. J. Clin. Invest. 61: 624–632, 1978.

23. Eggleton, M.G. The diuretic action of alcohol in man. J. Physiol. 101: 172–191, 1942.

24. Van Dyke, H.B. and Ames, R.G. Alcohol diuresis. Acta Endocrinol. 7: 110–121, 1951.

25. Fuchs, A.-R. and Fuchs, F. Alcohol effect on oxytocin release and uterine motility. In Biochemical and Clinical Aspects of Alcohol Metabolism, Sardesai, V.M., ed. Chas. C. Thomas, Springfield IL, 1969. pp. 105–114.

26. Zlatnik, F.J. and Fuchs, F. A controlled study of ethanol in threatened premature labor. Am. J. Obstet. Gynecol. 112: 610–612, 1972.

27. Ethanol in the treatment of threatened premature labor. Med. Lett. Drugs Ther. 18: 42–43, 1976.

28. de Wied, D. Peptides and behavior. Life Sci. 20: 195–204, 1977.
29. Hoffman, P.L., Ritzmann, R.F., Walter, R. and Tabakoff, B. Arginine vaso-
    pressin maintains ethanol tolerance. Nature 276: 614–616, 1978.

Review

Cicero, T.J. Neuroendocrinological effects of alcohol. Ann. Rev. Med. 32: 123–
142, 1981.

13. A Note on Fetal Alcohol Syndrome

Recently there has come to light in the medical literature a fact that has been known to folklore and to art for centuries: children born to alcoholic mothers are often abnormal. This syndrome has recently become the subject of a great deal of interest and research effort. The field is changing too fast to warrant a summary in this book, and an in-depth documentation of the latest data would soon become outdated. Instead, I will highlight the significant findings and refer the reader to some of the many excellent recent reviews.

Clinically, the effects of maternal heavy drinking are recognized by a syndrome of fairly specific facial abnormalities, small head circumference, and low body weight, occasionally combined with other malformations. The facial characteristics of the fetal alcohol syndrome are easily recognized by experienced physicians. The full-blown syndrome may appear only as a result of quite heavy maternal drinking and may be infrequent. More commonly, at lower levels of alcohol consumption there must be subtler effects on bodily and mental growth. Early doubts about the existence and prevalence of the fetal alcohol syndrome as such should not obscure the now evident fact that ethanol does indeed damage the fetus.

Clinically oriented reviews include those by Streissguth et al. [1] and by Clarren and Smith [2]. An excellent critical review by Abel [3] covers clinical and animal research and points out some of the difficulties in controlling experimentation with animals.

Ouelette et al. [4] reported a major clinical study where abnormalities were found to correlate roughly with the extent of alcohol intake, according to histories. Congenital malformations were found in 32% of the infants of the heaviest-drinking group of mothers. Poor growth and reduced head circumference were also more common in this group than among the offspring of lighter drinkers. The effects of smoking were not analyzed here, which is a drawback because tobacco is a suspected teratogen and its use correlates strongly with that of alcohol. Another clinical study [5] dealt with children up to 21 years of age, whose mothers had been reported to be alcoholic. Most of these children had a head circumference more than two standard deviations below the normal mean. Furthermore, their mean IQ was only 65. This reflects the most predictable and damaging effect of maternal drinking and suggests that low levels of drinking may have deleterious effects on intelligence and behavior that would be difficult to detect at birth.

There are few reports of withdrawal reactions in neonates whose mothers had been drinking just before delivery. This may be because the infant's liver is unable to metabolize ethanol, and withdrawal actually occurs slowly.

In human subjects, alcohol drinking histories provide unreliable estimates of the dose and timing of alcohol intake. Furthermore, many other factors, such as smoking, malnutrition, and poor hygiene, accompany abuse of ethanol. These can be eliminated in animal experiments, although nutrition continues to be the most difficult variable to control, as in all experiments with chronic ethanol administration. Several species have been studied, including rats [6], mice [7,8], miniature swine [9], and beagles [10], and a variety of effects have been demonstrated in fetuses carried by alcohol-treated dams. Some studies have shown no damage at moderate doses and blood levels, but more often the birth weights and litter sizes are decreased. Postnatal growth is usually retarded. Gross malformations are rare in these experiments, but do occur in some studies in a dose-related way. Most disturbing is the reported disruption of differentiation of the brain. Studies of the effects of alcohol intake during short periods of gestation are just beginning. Overall, there seems to be a general effect on growth that persists throughout pregnancy and may affect the brain at any stage, but malformations of single organs are apparently relatively rare.

Since ethanol disorders cell membrane structure (Chapter 4), it is easy to imagine how it might disrupt the elaborate and orderly steps in cell division and differentiation. In fact, it is surprising that more ill effects have not been reported. One might imagine that even brief episodes of high ethanol concentration could interrupt some critical developmental process.

Common sense suggests that this does not happen often, since many women drink occasionally and most infants are born healthy. The fetus is protected from many adverse influences (maternal malnutrition, for example) and may have ways of getting through brief periods of intoxication. Nevertheless, we do not really know the effects of small amounts of drinking during pregnancy. We lack evidence about dose relations. We do not know whether there is a threshold dose or whether there are critical periods of special vulnerability during gestation. At this time it is difficult to know what advice to give women who are pregnant or who may become pregnant. If we assume that some damage may be done in the earliest weeks of gestation before the woman knows she is pregnant, then we must extend our advice to all women of childbearing age. Although it might sound sensible to counsel them to refrain entirely from drinking (and although this would surely solve the problem, if the advice were followed), this position is extreme. Is it justified? Research proceeds so rapidly now on the effects of ethanol on fetal development that we may hope to have some clearer answers in the foreseeable future.

Summary

Ethanol intake during pregnancy certainly has adverse effects on the offspring, including retardation of physical and mental development. Malformations are less predictable, but they do occur. Both slow growth and malformations have been shown in animal models free of the confounding variables that exist in human populations. The question now is not about the existence of the syndrome but about its dose relation and whether there is a threshold for safe drinking for the women at risk.

References

1. Streissguth, A.P., Landesman-Dwyer, S., Martin, J.C. and Smith, D.W. Teratogenic effects of alcohol in humans and laboratory animals. Science 209: 353–361, 1980.
2. Clarren, S.K. and Smith, D.W. The fetal alcohol syndrome. New Eng. J. Med. 298: 1063–1067, 1978.
3. Abel, E.L. Fetal alcohol syndrome: behavioral teratology. Psychol. Bull. 87: 29–50, 1980.
4. Ouelette, E.M., Rosett, H.L., Rosman, N.P. and Weiner, L. Adverse effects on offspring of maternal alcohol abuse during pregnancy. New Eng. J. Med. 297: 528–530, 1977.

5. Steissguth, A.P., Herman, C.S. and Smith, D.W. Intelligence, behavior, and dysmorphogenesis in the fetal alcohol syndrome: a report on 20 patients. J. Pediat. 92: 363–367, 1978.
6. Abel, E.L. and Dintcheff, B.A. Effects of prenatal alcohol exposure on growth and development in rats. J. Pharmacol. Exp. Ther. 207: 916–921, 1978.
7. Chernoff, G.F. The fetal alcohol syndrome in mice: an animal model. Teratology 15: 223–229, 1977.
8. Randall, C.L. and Taylor, W.J. Prenatal ethanol exposure in mice: teratogenic effects. Teratology 19: 305–311, 1979.
9. Dexter, J.D., Tumbleson, M.E., Decker, J.D. and Middleton, C.C. Fetal alcohol syndrome in Sinclair(S-1) miniature swine. Alcohol. Clin. Exp. Res. 4: 146–151, 1980.
10. Ellis, F.W. and Pick, J.R. An animal model of the fetal alcohol syndrome in beagles. Alcohol. Clin. Exp. Res. 4: 123–134, 1980.

Index